MINERAL RESOURCES IN REGIONAL AND STRATEGIC PLANNING

This book is to be returned on
or before the date stamped below

Mineral Resources in Regional and Strategic Planning

PETER W. ROBERTS
Lanchester Polytechnic,
Coventry
and
TIM SHAW
The University of Newcastle,
Newcastle upon Tyne

Gower

© Peter W. Roberts and Tim Shaw 1982

Published by
Gower Publishing Co. Ltd.
Gower House, Croft Road, Aldershot, Hants., England

British Library Cataloguing in Publication Data

Roberts, Peter W.
  Mineral resources in regional and strategic
  planning.
  1. Mines and mineral resources
  I. Title II. Shaw, Tim
  333.8   HD9506

  ISBN 0-566-00395-3 ✓

Reproduced from copy supplied
printed and bound in Great Britain
by Billing and Sons Limited and Kemp Hall Bindery
Guildford, London, Oxford, Worcester.

Contents

SECTION I

THE ORIGINS OF CONCERN

1 Introductory Perspectives

During the past two decades concern has been expressed by all levels of government in the United Kingdom regarding the development of indigenous mineral resources. Increases in world commodity prices, shortages of key minerals such as oil and the need to ensure a continuous provision of aggregate materials have stimulated, not a planned response, but rather a series of pragmatic measures. These measures have been advanced by the United Kingdom government in order to ensure the supply of essential resources at a given point in time, with little regard for the long term strategic consequences. In the early 1970's the granting, by central government, of licences for a range of mineral exploration surveys created widespread concern in the areas affected by such programmes, although subsequently few positive developments have ensued from the initial exploratory investigations. Other independent policy decisions have been made regarding the development of North Sea oil and gas, deep mined coal, potash and mineral aggregates.

The present authors, through their association with the North West Petrochemicals Working Party (which was responsible for considering a wide range of strategic issues related to oil, gas and chemical industry developments), their familiarity with the Regional Aggregates Working Parties and their own work on offshore and onshore oil and gas developments in the North East of Scotland and North West England, became aware of the need for a comprehensive and co-ordinated approach to the strategic planning of minerals. This recognition of the need for a thorough examination of the role of minerals, and the lack of national and regional planning procedures for dealing with the problems generated by mineral extraction, was reinforced by the work of the Stevens (1976) and Verney (1976) Committees, and by the increasing controversy that accompanied proposed mineral developments during the late 1970's.

The authors are not of the belief that planning decisions, once made, should be inviolate; clearly changing circumstances will necessitate shifts in stance at local, regional and national levels. Indeed it could be argued that sensible planning strategies should include a number of future options based upon the central assumption that their robustness should be the ultimate aim. However, it is not suggested that there should be any support for a wholesale abandonment of policies which have been carefully formulated over a number of years in order to capitalise on a new opportunity without giving careful consideration to the implications of such change. For example, the Ross and Cromarty Development Plan was amended incrementally in order to accommodate the short term need of the offshore oil and gas industry. As experience has shown, the stability and gradual economic growth that had been previously brought to the Three Firths area of Ross and Cromarty was severely disrupted by the advent of land based construction work associated with offshore oil and gas. It is inevitable that the longer term consequences of such a construction boom will weigh heavily upon the people and local government in this part of the Scottish Highlands for many years to come.

This example illustrates a number of issues pertinent to the theme of this text. In the first instance it highlights the dangers of over reaction to new or changed circumstances. Had the oil boom not been allowed to burst so violently upon the economy and environment of North East Scotland it is conceivable that those activities which offered a longer term solution to the social and economic needs of such an area would not have been so damaged. The adverse impact upon industries such as British Aluminium at Invergordon and service industries for the whole of the area should not be underestimated. The haste with which oil associated developments were permitted in this area caused problems that could have been avoided or minimised. In a similar vein it should be noted that the reaction to oil and gas by central government, prompted by the views expressed by the oil industry, caused other related problems which could have been avoided had any reaction to the development opportunities been more considered. Contracts between government and the industry negotiated in haste arguably gave the industry too free a hand. In relation to the government's approach to depletion rates, significant volumes of oil and gas could have been left untapped, this allowing for slower depletion and the generation of more comprehensive and coherent planning strategies.

The desire to respond to the supposed needs of the expanding offshore industry resulted in the unnecessary over provision of facilities, such as the unused £14 million platform construction yard at Portavadie. The high capital costs of constructing these facilities together with associated social and economic infrastructure provision can be seen as unnecessary expenditure to be deducted from the profits of oil and gas; but the dashing of local expectation in places like Portavadie must have been considerable. It should surely be a tenet of sound planning that such wasted expenditure and effort should not occur. It is not a sufficient excuse by government to lay the blame for such waste at the door of speculative, private sector development, for both central and local government sanctioned these actions through their general attitude and statutory land use planning responsibilities.

The strategic importance of offshore oil and gas to both national and regional economies was not fully appreciated during the early phases of development. In the absence of any attempt at comprehensive forward planning, decisions were made on an incremental basis to serve the needs of an immediate situation. Neither of the major political parties seem to have been in much doubt as to the overall strategic importance of offshore hydrocarbon resources during the periods of their respective administrations over the past two decades, but whether for reasons of short sightedness or political dogma, a coherent and comprehensive strategy for the development of these resources, together with their associated activities, seems to have been largely neglected or rejected.

BROADER PERSPECTIVES

The authors are of the opinion that although other mineral resources may not have the same immediate economic significance as oil and gas, nevertheless they do play a vital and important role in national economic growth, industrial development, urban expansion and infrastructure provision. As such, there is a strong case to be argued for a measure of government intervention in ensuring the most efficient utilisation of a range of minerals, bearing in mind the legitimate interests of those involved in the winning and supply of minerals, those who require minerals for their industries and also those whose existing land use activities, recreation, amenity and investments are threatened by mineral extraction. It is clear that during the post war period, when there has been a massive increase in the demand for all manner of indigenous mineral resources, there have been significant areas of conflict between those who would develop such wealth and those determined to resist. What is also clear is that during the post war period there has been little in the way of a consistent framework in which to rationally consider all the various interests associated with the development of minerals. It cannot be denied that local authorities have included minerals in Development Plans (including structure plans) but essentially all applications for mineral development, whether considered purely by local government of referred to the Secretary of State, have been considered strictly on their own merits. Even though some applications, rejected by local government but sanctioned in Whitehall, have been deemed to be in the national interest, this can hardly be considered as a substitute for a strategy for mineral development. Although the Stevens Committee (1976) were not in favour of a strategy for each individual mineral, neither were they in favour of adopting a totally pragmatic approach to mineral development where local government and the minerals industry were without guidance and guidelines upon which to base judgements and decisions.

Although governments of different political persuasions have significantly divergent attitudes as to the way in which they should or should not intervene in the development process, and that by definition must include minerals winning, the post war period, following the publication of the Waters Committee Reports (1948 to 1955), the first and second editions of the Control of Mineral Working (1951, 1960) and a number of pertinent clauses within the various Town and Country Planning Acts, has been characterised by a largely mixed approach by government. Clearly there has been some government intervention, and those governments which have favoured less rather than more intervention in the workings of the economy, have themselves reserved 'call-in' powers should the dictates of the economy, strategic considerations and political expediency necessitate such action. Socialist governments have arguably been more determined to maintain greater control over the development process but conventional wisdom within the Labour party and in the post war period has favoured allegiance to the mixed economy and consequently the various components of the mineral industry have exerted a considerable influence on the pattern of mineral development in the United Kingdom. There have been those who have resisted and resented government intervention, even at the local level, but equally many have accepted the need for some measure of control or at least have become resigned to the fact that most political parties will seek, through legislation, the power to exert some control over issues related to mineral extraction.

If successive governments of the post war period in the United Kingdom have retained their powers to intervene in the decision making process as far as mineral resources are concerned, it might be reasonably asked why therefore it is felt that a need exists to further investigate this field of planning for minerals. The response to such a question is threefold: the doubt as to whether governments fully appreciate the strategic significance of minerals and the physical, economic and social implications of permitting their development; doubt about the effectiveness of the planning and intervention powers that government retains and doubts about the ways in which such powers are used to ensure an efficient minerals industry without long term social and environmental costs.

It might well be argued that one of the main difficulties encountered in the history of post war planning for minerals is that there has been a lack of a coordinated and comprehensive strategic approach to mineral development. If, as Stevens (1976) would seem to indicate a national strategy for all minerals is not appropriate, then there is certainly a strong case to be advanced in favour of a set of guidelines which identify the sort of approach which should be adopted by central and local government planners. Approaches similar to those adopted in the Scottish Development Department's National Planning Guidelines, Aggregate Working (1977), might be deemed appropriate for application to minerals throughout the United Kingdom. Even so there is clear evidence from earlier national guidelines prepared by the Scottish Development Department, for example the North Sea Oil and Gas Coastal Planning Guidelines (1974), that preferred strategies have not always been adhered to in the face of demands from developers. The danger that exists in all such proposed guidelines is that if they are produced in isolation from each other then inconsistencies and contradictions can be built into any planning approach.

There have quite clearly been interesting attempts made to produce planning strategies for mineral development that aim to maximise the benefits to society through the equitable distribution of profits, but with a minimisation of economic, social and environmental disruption. The Zetland County Council Act (1974) was a case in point but it should be appreciated that this was probably easier to achieve in a remote island situation with a less complex economic structure and located in a key strategic position vis a vis a particular mineral resource.

Not all legislation has achieved what many would have hoped for in terms of controlling the impact of mineral associated developments. In the case of the

National Parks and Access to the Countryside Act (1949), together with other supportive legislation of the post war period, one of the twin aims was to 'preserve beautiful countryside from unplanned and inappropriate development' (Sandford, 1974). It is evident in the Peak District National Park there has been continuing despoliation of landscape through limestone quarrying. Landscapes on the fringe of the Park have also been spoilt and operations now threaten to advance into the Park area itself (Harrison and Machin, 1981). Associated problems of noise, dust and traffic generated through quarrying exert an adverse effect upon the amenity and aesthetic qualities of the National Park. Clearly, there is incompatibility between the aims of National Park planning and economic planning. Even the most dedicated conservationist would be forced to conclude that a strict adherence to National Park principles would cause a loss of jobs and shortfalls in the production of important raw materials for industry and construction. The dilemma for strategic planning is evident; the balancing of regional and national interests between amenity, recreation and conservation on the one hand and jobs and economic growth on the other. This problem of conflicting government policies is one that has faced all governments in the post war period where support has been given to the broader aims of conservation and to the need for minerals which are basic to the achievement of economic growth.

COORDINATED APPROACHES

It is not the intention or expectancy of this text to produce a universal panacea for the ills that beset minerals planning, but rather the book aims to identify a more rational and coordinated approach to planning for mineral development, giving due regard to important regional and strategic planning considerations. The emphasis given to strategic and regional planning issues will require justification, but from the basis of prior knowledge and experience there is much to suggest that the objectives of local communities, district planners and mineral operators are not always consistent with the broader objectives of national and regional planning.

Given that many of the decisions surrounding minerals winning are taken by local authority planning departments operating under the guidance of draft or approved structure plans, it is likely that there will be variations in attitude to issues related to the development of minerals. Although in reviewing submitted structure plans and considering the evidence presented at their Examination in Public, the Secretary of State for the Environment is in a position to recommend amendments that would provide a more appropriate context for minerals development, this is rather a cumbersome and inefficient method of coordination. Given the delays that have occurred in the submission of structure plans, differences in approach to their preparation and the prospect of them being reviewed within a relatively short period of time, there is little encouragement for planning authorities to promote a favourable climate within which sensible forward planning, by government and the minerals industry can occur.

It is worth considering the contribution of the Scottish Regional Reports (although few have considered minerals), for within these Reports broader and more sensible short and long term perspectives have been considered. Coordination, via the Scottish Development Department, of a relatively small number of Scottish Regional authorities would appear to create fewer problems than attempting to coordinate linked issues emanating from a substantial number of county structure plans. There is every indication that government feels there should be some form of coordination at a regional level. This is demonstrated in the establishment of regional working parties to examine the current state and future pattern of the mineral aggregates industry.

Evidence to date would seem to indicate that the activities of these regional working parties have been variable with omissions and inconsistencies in the data made available. The lack of annual monitoring from some of the most important aggregate producing and consuming regions could reduce the effectiveness of regional and strategic planning and decision making. However, in looking for better approaches to minerals planning, the contribution of such regional working parties cannot be rejected though there is a need for them to be more consistent and comprehensive in their data collection and its subsequent analysis.

Coordination is seen as an important theme within this book. In a similar light to their plea for greater coordination between those working on minerals issues, within and between regions, the present authors would suggest that the formulation of policy by central government departments should aim to achieve a greater measure of consistency. This is needed in order to minimise the problems encountered by those working in, for example, National Parks where one element of regional and national strategy clashes directly with another.

Although, for reasons which hopefully will become clear, it is felt that as far as minerals planning is concerned there is a need to develop a regional approach; this is not to dismiss particular local issues or to subjugate all local attitudes to the regional plan. Rather it is to ensure that the best use is made of available mineral resources. Indeed, it could be argued that many areas would suffer less from the impacts of mineral working if there was the discipline of a regional plan which provided a clear framework for the industry. Clearly, many of the issues developed in the context of a regional approach to minerals planning do have a bearing upon individual locations and pertinent local issues therefore have to be discussed.

If industry is to prosper then at some stage mineral resources have to be developed. The impact of mineral extraction is considerable, though this will vary according to the type of mineral involved and the extent of the reserves. Substantial opposition is generally associated with proposals to exploit mineral

wealth, although the scale and intensity of such opposition will be varied.
For example, on amenity grounds, there was well orchestrated opposition to the
proposals of Rio Tinto Zinc to mine metaliferous ores in Snowdonia National Park
(Smith, 1975). Interference with other economic activities such as agriculture
usually generates hostility between the farming community and the National Coal
Board as at the recent Vale of Belvoir Inquiry or the National Coal Board's
Opencast Executive in the case of the large opencast coal mining project at
Butterwell near Morpeth in Northumberland. The varied sources of objection to
proposed minerals developments highlight the needs for society, through
government, to decide upon its priorities. Although in a broad sense most people
benefit from the free flow of minerals to industry, a decision has to be made as
to the price society is prepared to pay for any restrictions, for whatever reason,
placed upon the movement of minerals. Although hypocracy is perhaps too
strong a word to use, clearly there are many individuals, organisations and
government bodies who, whilst accepting the need for mineral development,
would rather it took place away from their own areas and areas of interest; this
'anywhere else but here' syndrome is well described by Dean (1974).
Government would be wise to generate a more sensible atmosphere in which
forward planning and development can take place. The need for such an
approach can be seen in recent public inquiries involving minerals issues. The
Vale of Belvoir public inquiry demonstrated conflict not only between the
National Coal Board and pressure groups, but also between individual government
departments. The Countryside Commission, a government financed body, was
also in opposition to the proposals of the National Coal Board. Presumably the
conflict between these bodies could have been lessened had government decided
in advance that there were national priorities. If coal production is vital to the
country's future industrial and economic performance then such energy
requirements should in such circumstances perhaps override agriculture and
amenity considerations. Public debate, frequently costly, could then be
focussed on the ways in which agriculture and amenity can be least disrupted,
rather than on whether these interests should prevent new coal mining from taking
place at all. Although it is a little unfair to prejudge a response from the
Secretary of State to the Inspector's report, it is difficult to imagine the plans
of the National Coal Board being thwarted.

CONTENT AND STRUCTURE

It is not the intention of this book to concentrate on the historical elements
and current practice in minerals planning and the extractive industries. It
introduces these themes as a basis for assessing the need for change and the form
that such change might take. Thus this book has only a short chapter
highlighting one or two significant milestones from the past and then moves to a
consideration of the importance of minerals to the economy and the effects of
minerals upon regions. However, due consideration is given to the

contentious issues that surround mineral working and the longer term consequences of past and present activities. The third section of this book is concerned primarily with decision making, evaluation and planning. Within this section an attempt is made to understand the reasons why planning legislation has progressed to its current state and the implications this has had for the various sectors of the minerals winning industry. The final section is directed to looking at the raison d'etre for, and methods of achieving, a coordinated and comprehensive approach to strategic minerals planning mainly at a regional level.

Obviously the main thrust of this book is directed towards the arguments for the adoption of a coordinated and comprehensive approach to minerals planning and it is appreciated that there is a need to take account of many more issues than minerals alone. To this end it attempts to take account of new approaches to evaluation which, by definition, purport to be comprehensive. For example, the much vaunted Environmental Impact approach is considered as a means of evaluating the range of social, economic and physical environmental consequences coincident with various forms of mineral extraction. Environmental Impact Assessment is important because it can promote understanding and provide a foundation for modifying or reformulating regional planning strategies for mineral development. At all times the present authors have attempted to retain a perspective of realism and are aware of the dangers of becoming too dependent upon particular tools of evaluation, such as cost benefit analysis, which have influenced decision making during the past twenty years and which have been found wanting in many instances. (Clifford, 1978).

Although it is unlikely that a simple transition can be made from what is seen as an inadequate minerals planning system, to a more coordinated and comprehensive regional and strategic approach, the present authors are anxious to identify the seeming advantages of such a change. The passage towards comprehensive approaches is not likely to be an easy one. However, it surely bodes ill for the future of minerals planning in the United Kingdom, and probably for other nations as well, if governments and their various planning agencies are not prepared to grasp the nettle and strive to find more acceptable pathways through the maze of vested interest, pragmatic responses and confused and conflicting government policies.

Although it was never the intention of the present authors to become overtly politically biased in their writing, and as the book reveals, criticism and praise have been directed at contrasting political influences on planning, it is important to give consideration to the relationship between different governments and the minerals industry. There can be little doubt that the minerals industries, whether privately owned or nationalised, seek to exert a high degree of influence upon the decision making process of all governments and thus influence the way in which mineral extraction develops. Any attempt to justify the need for a coordinated and comprehensive regional approach to minerals planning, cannot afford to overlook patterns of ownership, control and organisation of minerals winning operations and the related needs of industry. The era of the company mine and the associated town, which characterised the industrial revolution of the nineteenth and early twentieth centuries, left its mark in the form of isolated, despoiled and redundant communities. Although modern mining companies are now responsible and responsive to the broader needs of society they cannot be left, by default, to make planning decisions. As will be demonstrated later in this book, many regional economies are highly dependent upon the winning of minerals. It is the view of the present authors that planning in such a context must anticipate the demand for minerals and the demands of the minerals industry rather than responding to them in an incremental fashion.

ACKNOWLEDGEMENTS

The authors acknowledge the assistance of many people in the preparation of this book. Some of the fundamental ideas that are developed in this text have their origins in the work of the Celtic and Irish Sea Oil Research Unit set up in the Department of Town and Country Planning at Liverpool Polytechnic and directed by the authors between 1976 and 1978. Gratitude is expressed to Liverpool Polytechnic and to Merseyside County Council whose financial support and encouragement made such work possible. In a similar vein, the assistance of the Department of Urban and Regional Planning at Coventry (Lanchester) Polytechnic and the Department of Town and Country Planning at the University of Newcastle upon Tyne is noted, for without the support of their respective institutions the authors would have found it difficult to build upon the foundations laid down in Liverpool. Thanks are also due to colleagues in both the academic and practice worlds who have provided much useful help, advice and criticism.

Our gratitude is also due to Mrs. G.E.Davies of the Department of Town and Country Planning at the University of Newcastle upon Tyne who has transformed an appallingly written manuscript into a presentable typed document.

Not least of all our thanks to our wives who fed us, tolerated us and proof read draft manuscripts during the months of preparation.

2 Developments Towards Minerals Planning

Although there was virtually no planning control over mineral working in the period before the Second World War, this did not mean that minerals winning operations had not played an important role in the social and economic development of the British Isles. Rather it was a reflection of the way in which industry, government power and society as a whole was organised. It was also a reflection of the way in which the minerals industry was perceived. Minerals were seen as being vital to the economic growth and future well being of a mature manufacturing nation and as such the exploitation of these resources was deemed to be of paramount importance. In short, it was easy to see the obvious advantages of mineral extraction but perhaps less so to appreciate the problems which were associated with the increasing scale and extent of mineral developments in Britain throughout the 19th and early parts of the 20th centuries.

Planning control over mineral extraction is therefore very much a phenomenon of the post war period, but as the foundations of current land use planning were laid down immediately after the second World War with the enactment of the 1947 Town and Country Planning Legislation it is clear that some of the major influences which gave rise to this Act must have been active for some considerable time before the war. It is not the intention here to embark upon a detailed social and economic analysis of the development of industrial nations, but rather to attempt to establish the role that minerals played in the process of industrialisation and to identify the growing areas of concern which eventually lead to mineral developments becoming controlled by land use planning law.

In embarking upon any discussion of minerals and minerals planning it would seem important to establish some definition of minerals. Minerals can be an emotive word conjuring up, perhaps, images of precious gem stones and metals. To develop a categorisation and definition of minerals in such a restricted context would be inaccurate, misleading and singularly inappropriate when pursuing a theme of planning for mineral development. Therefore, it would seem far more appropriate to adopt a definition for minerals which encompassed all mineral resources, particularly as it is the more commonplace ones which have contributed so greatly to growth in industrial societies, and the development of which still pose problems to planners.

Definitions of minerals vary from the precise scientific; 'a natural inorganic substance which possesses a definite chemical composition and definite physical and chemical properties' (Moore, 1963) to the more general; 'Occurring naturally in the earth' (Uvarov, 1956). It is also worth noting the definition which was included in the 1947 Town and Country Planning Act (Section 119(1)). Here minerals were defined as including;

> 'all minerals and substances in or under land of any kind ordinarily worked for removal by underground or by surface working: provided that it shall not include peat cut for purposes other than sale'.

In this book the wider view of minerals is taken to include the largely organic hydrocarbons and the composite rock minerals in addition to metaliferous mineral resources. Moving on from a broad definition of minerals, the authors have adopted in this text the system developed by Blunden (1975) for categorising mineral resources.

Blunden's approach is straightforward dividing minerals firstly into 'metaliferous' and 'non metaliferous' and secondly into 'ubiquitous' and 'localised'. He has further refined his categorisation by introducing economic criteria. Although it is not the intention here to use the five groups that Blunden suggests to value minerals, it is recognised that various economic criteria must be taken into account when considering the development of these resources in their historic, current and future contexts.

ROLE IN ECONOMIC GROWTH

Throughout history minerals have had an influence on the evolution of societies and civilisations. The impact of the discovery of metaliferous ores together with the development of a technology capable of processing them was significant (Darby, 1956). Although the ability to process and work with stone and metal represented major staging posts in the development of civilisations all over the world, the identification and use of other minerals throughout history has helped to further the development and prosperity of societies in many countries. For example, the use of fuller's earth was important to the development of the woollen industry in the Middle Ages, an industry which brought great wealth to the nation (Clapham, 1963), and of course coal became the primary fuel for the Industrial Revolution in the critical development from water power to steam.

Spooner (1981), has referred to minerals as having a 'place value'. It is clear from something as commonplace as the variation in vernacular architecture throughout the British Isles that in different locations there has been a response to the mineral resources available for building; limestone in the Peak District, gritstones in the high Pennines and slate in Snowdonia. However, whilst all areas have managed through time to adopt local materials for building and road making, not all areas have been blessed with mineral wealth appropriate to the needs of industrialisation, and it is the extent of the demand for certain categories of minerals that has caused and continues to cause a variety of social, economic and environmental problems in areas of mineral extraction. This has also caused high levels of both international and inter regional transfers of minerals.

Because of the comparative ease of transporting finished products, given their higher value per unit load, there has been a tendency for industry to be attracted to the areas where the basic minerals upon which industry is based are mined

(Lloyd and Dicken, 1977).   This was very much the case when communications were poor (a point developed at a later stage) and it would have been very difficult to transport the minerals over any great distance.   As coal became such an important mineral in the process of industrialisation it was not surprising to see heavy industry and associated manufacturing developing in coalfield areas.   Not only is it evident that minerals were vital to industrial development, but also that they were significant in the process of urbanisation in what are now the old industrial areas of the western world (Warren, 1973).

Wherever minerals are developed they will have a range of impacts upon those areas.   The scale of operation is an important consideration here.   In earlier centuries the size of operations was small, either because of the limited extent of the reserves, or because of the technological constraints placed upon mining. Given the passage of time, many of these early workings have disappeared. Because they were small and often short lived they never gave rise to any appreciable scale of urbanisation.   As the scale of mining operations became larger, as a result of technological advancement and an increasing domestic and world market for both mineral products and manufactured goods, so too the impact of mining itself and the industry to which it gave rise increased.

Even with dramatic improvements in transportation during the past century there has still been a tendency for industry to be drawn to many of the locations from which their basic minerals are derived.   Clearly, as time has progressed the factor of industrial inertia has been an important consideration.   Although coal was transported all over the world for the purposes of domestic consumption and power generation it was still an expensive commodity to transport and there was no economic reason to move coal to industry when industry could more readily develop on the coalfields.   This was certainly the case in countless examples in the 19th century and the early parts of the 20th century.

Coal was not always the prime attraction however, as demonstrated by the development of the iron and steel industry in Leicestershire and Northamptonshire which was based upon Jurassic iron ores occurring locally.   Poor quality ores and the distance from coal, not to mention competition from the West Midlands, had an inhibiting effect upon the development of the area as a major steel producing region but changes in technology ultimately allowed a part of the area, particularly Corby, to become an important steel manufacturing town (Manners, 1972).   This was clearly a case of a new mineral resource giving rise to industrial development in situ.   Brick making was another example where, in part, it was the presence of suitable clays which gave rise to adjacent brick works. Although in some cases the clays used were found as part of the coal measure series and as such there was a fortuitous juxtaposition between the clay and coal for firing, a large portion of the country's brick making capacity was located close to Bedfordshire and Peterborough on Oxford Clay.   These sites were a long way from coal fields but evidently it remained cheaper to carry the requisite amount of coal to the brickworks than clay to the coalfields. By chance such sites provided mid point locations between the source of the fuel and the market for the final product (Lloyd and Dicken, 1977).

The importance of minerals to economic growth is in part determined by the cost of their extraction and movement to a market. Indeed, it has been frequently acknowledged that minerals only take on the mantle of a viable resource when there is a demand for them and when technology allows such minerals to be extracted and then if only at an acceptable cost. Zimmerman (1951) suggested that 'Resources are not, they become'. In any consideration of minerals due account should be taken of transport factors. Certain minerals have been more capable of withstanding the costs of transportation than others. Metaliferous ores have, generally speaking, been economically more viable to transport over long distances than many of the non metaliferous, low value, high bulk minerals such as sands and gravels.

Developments in transportation gradually meant that minerals could be moved more effectively (and within economic constraints) both within and between regions and countries. For example, coal mined by the side of the Tyne was taken to the growing domestic market in London from Elizabethan times. The cheapness of this form of transport for bulk goods, particularly as the size and reliability of ships increased, is reflected in the fact that long after the development of a comprehensive rail link between the North East of England and the South East, a substantial volume of coal was still despatched by sea. The tremendous development in naval engineering around the turn of the 19th century paved the way for new generations of larger bulk carriers capable of bringing minerals from all over the world to the industrial heartlands of Western Europe. Although partly due to the cheapness of production costs in other parts of the world, 20th century shipping has made it feasible to transport coal from as far afield as Australia and North America to Britain.

The development of an extensive rail network within Britain greatly facilitated the movement of minerals. Even relatively short hauls such as coal from the Yorkshire coal fields to the heavy and manufacturing industries of the West Riding and South Yorkshire were undertaken by rail. Lines were also built specifically to move mined and quarried minerals to the main rail network. It was not just in Britain that railways were important. Large mineral reserves located deep in the heartlands of continental land masses could not possibly have been opened up had it not been for the advances in railway engineering. Although with the increase in labour costs, railways are now very costly to build, in some cases if minerals are to be developed a new railway may remain the only viable way of realising the potential of a mineral rich hinterland. In 1949 high grade iron ores were identified in the north east parts of Laurentia in Canada and it was decided to build a new railway over very difficult terrain from Schefferville to Sept Isles on the St.Lawrence River (Watson, 1963).

Road transport for the movement of minerals was limited until there had been noteable improvements in engine and lorry building, which occurred particularly after the first world war. It was only with the reduction of the rail network that greater attention was paid to the development of an improved trunk road network; concurrently the size and capacity of lorries was increased. This was a feature of the post war period.

Canals and rivers have been used as a means of transporting minerals but in Britain their use as commercial waterways was to a large extent truncated by the more rapid and more extensive development of the railway system during the Industrial Revolution. The limited capacity of many of the inland waterways in Britain meant that they were incapable of responding to the demands of a rapidly expanding industrial nation. However, it should not be forgotten that some canals were built expressly for the purpose of carrying minerals. The Worsley Canal linked the Duke of Bridgewater's collieries with Manchester and there was clearly a lot of use made of newly constructed canals in the West Midlands and the Potteries in the 18th century for the movement of bulky mineral products (Hoskins, 1955). In other parts of the world inland waterways have been more extensively used for the transportation of mineral products, for example, the Great Lakes and St. Lawrence Seaway in North America. It is of course the ability of these waterways to carry large vessels which keeps them in use even in the second half of the 20th century.

It is clear then that technological developments have facilitated mineral working and transportation. Scientific and technological advances allowed the process of industrialisation and economic growth to move forward at a faster rate than in earlier centuries. It is therefore worth highlighting some of the important developments which allowed minerals to play a greater part in economic development.

The development of coal mining was hindered by the inability to extract coal other than where it outcropped. Surface and adit mining were possible but it was difficult to extend these workings. The inhibiting factors of tunnel collapse and faulted seams were significant. Even where shallow pits were dug the danger of collapse was always present and this gave rise in some areas to the development of 'bell pits' (Bracegirdle, 1973). A shaft was sunk and the coal mined in a small gallery beneath. Judgement and perhaps an element of luck dictated how large a gallery could be created through mining the coal before it collapsed. This was particularly inefficient in that it meant more time had to be devoted to sinking shafts, and valuable coal had to be left untapped. Throughout the 19th century progress was made in improving mining technology which allowed deeper mines to be sunk (also here benefiting from the advances being made in geological survey and mapping) and larger subterranean seams to be worked. Due to the low level of prevailing technology, manpower was used to carry the coal to the surface. This was highly inefficient, despite at times the seeming expendibility of the human resources involved (Chadwick,1842). Therefore, developments in mechanised lifting were heralded as a major development in deep mining operations.

There was an important development in steel making in 1878 following the perfection of what became known as the Gilchrist-Thomas process. This allowed ores with a high phosphoric content to be used in the making of steel. Although there were still doubts about the quality of steel made by this process, in the 1930's Stewarts and Lloyds built a large steel making plant at Corby.

The case of the Northamptonshire Jurassic iron ore field serves to illustrate another aspect of technological innovation. Larger industrial complexes required supplies of minerals at competitive prices and the ability to work extensive areas of mineral ore at cheap rates was clearly a bonus. In Northamptonshire in the inter war period mechanised draglines were introduced. These were often rail mounted to make them more manouverable. This development meant that not only was it possible to extract the ore quickly and cheaply, but it was also possible to remove the overburden in a similar manner. In looking forward to some of the causes for concern which contributed towards the introduction of land use planning controls for mineral developments, it was the increasing scale of new mineral winning operations in areas which previously had not been considered industrial, that made pressure groups and government bodies wonder how extensive, intrusive and damaging they would become.

It is perhaps worth touching very briefly upon the impact of war on mineral operations. It would of course be very difficult to gauge precisely which of the various war time innovations in the fields of science and technology eventually had either a direct or indirect bearing upon the minerals industry. Clearly war time shortages and the dramatic reduction in supplies of foreign resources resulted in a reappraisal of mineral reserves which had previously been abandoned for reasons of difficulty of working or economic viability in peace time conditions. Home based mineral production was thus boosted to meet the changing requirements of industries geared to war.

CAUSES FOR CONCERN

Much of the 19th and early 20th century concern about mineral winning activities was directed at issues of the health and safety of those who worked in the industry. The conditions which faced workers in the coal industry were dire, particularly for the large numbers of women and children involved. This was a feature common to many aspects of industrial employment in the 19th century where the conditions were hazardous, the hours long and the wages low (Roberts, 1948). In the second half of the 19th century a reform movement gradually gained momentum which was to force a range of social reform and

certainly improve the conditions in industry and mining.   To this movement
was gradually added the weight of the growing Trade Union movement.
Although such social reform might seem somewhat removed from land use planning
issues, it does serve to demonstrate how a growing concern about the organisation
of industry and society, once perceived and articulated, could be used to bring
about change through pressure and ultimately legislation.

In the last century the concern was primarily for people but little by little,
particularly as the scale of mineral winning operations increased, concern was
also expressed about the impact upon property, other land uses and eventually
upon aspects of aesthetics and amenity.   Mention has already been made of
the impact that new technology had upon the scale of certain mineral operations
and it is reasonable to assume that many mineral developments increased in size
as the process of industrialisation continued.

Damage resulting from mineral working was extensive, especially in areas of
deep mining.   Subsidence in coal and salt mining areas caused damage to
property and land alike.   The process of deep mining causes subsidence
(Wallwork, 1974) and although in the post war period there have been attempts
to reduce its extent they were not generally practised or in many instances known
about at the height of deep mining activities.   Even if these techniques had
been available earlier, it is doubtful whether they would have been adopted as
it would most certainly have reduced profit margins.

The loss of agricultural land to mining and quarrying was a cause for concern,
particularly in the inter war period.   In the 1920's agriculture was in a very
depressed state (Coppock, 1971) with a large acreage of land being willingly
sold to urban developers in a period of startling urban expansion.   To see
additional land being taken out of agriculture for mineral extraction was
therefore an additional source of concern for those who were worried about the
ability of the nation to feed itself, especially with the memory of food shortages
during the war still fresh in their minds.   The Report of the Committee on Land
Utilisation in Rural Areas (Scott,1942) was mindful of the impact that extractive
industries could have upon the nation's stock of agricultural land.

The same report also made mention of the harmful impact of some extractive
industries upon the appearance and the amenity of the countryside.   As long
ago as the days of Cobbet's Rural Rides there has been an element of concern
about the intrusion of urban and industrial developments into the countryside.
Literature of the past hundred years abounds in references to the appearance
of industrial development (Pocock, 1979).   More recent literature has not
been slow to highlight the impact of mining upon the landscape (for example,
 Orwell, 1920 to 1940) and this has been supported by a wealth of literature
from academic sources and amenity and conservation organisations all
stressing the impact of industry and related developments upon the general

appearance of the countryside. However forceful might have been the arguments in post war Britain for the need to extend planning control over industrial development in the interests of amenity and conservation, the conservation and amenity movement has its origins in the first half of this century, before the second world war. For example, Tunbridge (1981) demonstrates the increasing involvement of the National Trust in conservation in the inter war period. In a similar vein, Sandbach (1978) identifies the role of conservation bodies in pressing for the establishment of National Parks. Rural conservation in inter war Britain was gaining momentum and attracting the attention of both central and local government (Sheail, 1981).

LEGISLATION PRIOR TO 1947

It is important to stress that before the Town and Country Planning Act, 1947, there was very little effective planning control over minerals. However, there was some legislation which had a bearing upon the minerals industry which arose largely from the causes for concern which have already been introduced. Although not all such fears resulted in the introduction of legislation, prior to the second world war there were a number of pertinent reports commissioned by government which can be deemed to have influenced subsequent planning legislation.

The reform movement of the 19th century exerted pressure which helped to bring about the enactment of a range of legislation including the Mines and Collieries Act of 1842, which amongst other things prohibited the employment of women and girls underground and raised the age limit for boys to ten. A sequence of Education Acts brought an indirect reform to working in the minerals industry in that compulsory education for children was introduced and gradually the school leaving age raised. Although the 1842 Act had made some important progress in respect of improving safety and well being in the mining industry, subsequent measures were to follow which introduced more stringent inspection of mines and later quarries.

Some progress was made at the end of the 19th century in securing compensation for certain types of damage to property caused by subsidence, for example the Brine Pumping (Compensation for Subsidence) Act of 1891 though there was no equivalent for coal mining subsidence. It is of interest to note however that the Kennet Committee, which reported in 1939 on the problems associated with land restoration following opencast mining operations in Northamptonshire, suggested that those responsible for and profiting from the mining should pay a substantial part of the restoration cost, a point later supported by the Scott Committee (1942). There was, however, no move to bring mineral extraction under any form of land use planning control until the Town and Country Planning Act of 1932. This, together with the related Interim Development Order of 1946 are touched upon in a later chapter, suffice it here to say that there was little evidence of planning authorities exercising the powers given them under this Act.

Nevertheless, the seed had been sown and government had given notice that land use controls over mineral working would be introduced and enforced. The impact of conservation pressures had in part given rise to the Addison Committee, appointed in 1929 to enquire into the feasibility of setting up National Parks (Bracey, 1970). It was clear that organisations like the Council for the Preservation of Rural England were determined not to have large areas of scenic Britain fall prey to inappropriate development, including mineral extraction. Although these conservation groups were not solely interested in issues pertaining to minerals, it is important to realise that they were creating a greater awareness throughout the country about the threats to rural Britain. Subsequently this was to have an influence on the way in which both public and government approached the thorny question of mineral extraction.

Although there was a general appreciation of the nation's requirements for minerals and that this would clearly require a certain amount of extraction to take place in the countryside, there was concern that, as far as the more ubiquitous minerals were concerned, 'where deposits suitable for working are geographically widespread, areas which may be worked should be determined as far as possible nationally and, with the guidance of expert geological advice and in consultation with the industries concerned, should be incorporated into local planning schemes. In this way adequate utilisation of the country's mineral resources will be assured whilst the spoliation of the countryside by the casual opening up of quarries will be prevented' (Scott, 1942).

This contains a number of issues ranging from the need to conserve resources and the appearance of the countryside to the need to plan for resources through local planning but in the context of a strategic planning framework. The Scott Committee in their report had made much of the idea of a Central Planning Authority and it recommended that the aftermath of mining and quarrying operations should not be permitted to place a blight upon the countryside. They suggested that the Central Planning Authority should survey and evaluate the problems associated with dereliction and the means by which they could be restored and returned to some form of productive use. The undesirability of leaving land derelict was recognised even in the 1930's, when many public works schemes to occupy the unemployed were directed towards the clearance of despoiled areas. It could well be argued that it was not until central government took a much more positive approach to the problem of derelict land reclamation that significant progress was made in reclaiming the backlog of derelict land inherited from an earlier period.

Mineral wealth had created an environment for industrial growth in Britain for well over a century before the 1947 Town and Country Planning Act came into force. From a land use planning perspective it was clear that the vast majority of mining and quarrying activities had given little thought to the wider environmental implications of their actions. It was therefore to be the task of post war planning legislation to find an effective means of, and context for, controlling future minerals winning operations as well as dealing with the range of inherited problems from the period before appropriate land use controls were introduced.

SECTION II

SIGNIFICANCE, NEEDS AND CONFLICTS

3 Significance

The availability of mineral resources has, is, and will continue to be a prerequisite for the initial development and further growth of many national, regional and local economies. Indigenous mineral resources have, from the late eighteenth century onwards, provided the basis for the industrial growth of many regions and represent a rationale for settlement patterns and the distribution of population. Whereas, at a global level, the current consumption of mineral resources amounts to only 4.5 per cent of the total value of the world's economic output, (Cottrell, 1978) the significance of mineral resources is far greater than such a simple statistic suggests. In many areas the raw materials provided, directly and indirectly, from mining activities are basic to the regional economy. The provision of social and economic infrastructure is also dependant upon the availability of mineral materials. During the industrial revolution the spatial pattern of resource extraction and processing conditioned the form and function of settlements. The legacy of this resource based pattern of growth, today, provides planners with a number of severe and intractable problems.

Mineral resource development occurs in a cyclical fashion, from discovery, through exploration and exploitation, to decline. This pattern of development is closely linked to the supply and demand characteristics of a particular resource, and to the continually changing market for a resource. Notwithstanding the specific consequences of an individual resource development at a single location, certain general features and problems in the analysis of supply and demand for minerals can be identified.

Firstly, there is the difficulty of defining the extent, quality and quantity of any resource and thus, in estimating the available or potential level of output. Given changes in the definition of what constitutes a viable reserve, and taking into account the adjustments made through the coverage provided by detailed survey information, then the total recoverable amount of a mineral can vary in terms of its spatial distribution and through time. Prevailing market conditions also affect the viability of a particular source of a mineral, this factor reflects the interaction between supply and demand. An indication of the constant changes that occur in the evaluation of what constitutes viability, and thus, in the estimation of the size of the recoverable reserve of a mineral, can be illustrated by reference to the adjustments which have been made to the estimated size of the North Sea oil reserves.

Table 3·1 indicates the changing pattern of proven, probable and possible reserves in the United Kingdom sector of the North Sea.

Table 3·1

Estimated United Kingdom North Sea Oil Remaining
Recoverable Reserves - Existing Finds 1975 - 1980

Year	Proven	Probable	Possible	Total
1975	1,060	305	435	1,800
1976	1,350	580	360	2,290
1977	1,380	570	550	2,500
1978	1,405	625	590	2,620
1979	1,397	509	605	2,511
1980	1,200	625	575	2,400

Million tonnes
Source : Department of Energy (1975 - 1980)

Viability is also subject to variation due to changes in the level of technical
ability to develop a mineral resource, and because the commercial conditions
which control extraction alter through time.   Political and strategic considerations
also influence viability.   Therefore, for any geologically determined reserve, it
is possible to quantify the amount of a mineral which can be extracted at a given
state of technology, the amount which can be extracted at a commercial rate of
return, and the amount which it is desirable to extract given the need to, either
maintain a strategic reserve, or to achieve a target level of national self
sufficiency.

A second characteristic which must be considered is the relationship which
exists between the present and future patterns of demand for proven and viable
reserves of a mineral resource.   Even under conditions of overall market
stability, and assuming that the total level of demand remains constant, it is
incorrect to assume that the level of demand for a particular source of a mineral
will remain stable.   The market for aggregate minerals clearly demonstrates the
danger of assuming demand stability, for although overall regional demand may
not vary, the resource mix which is required can change, as can the sub regional
pattern of demand.   Such variations may reflect adjustments in demand due to
the requirements of specific projects, for example, the construction of motorways.
Transitory demand can cause adjustments to the sub regional pattern of supply.
Other variations in demand may reflect differences in the level of elasticity of
demand for a particular resource, and the degree to which substitute resources
can be provided.   The difficulties, which are encountered when attempting to
analyse the current level of demand for a particular mineral resource, are

exacerbated when attempting to forecast the future level of demand.

It is also important to recognise a third consideration, that is, that factors which are seemingly external to the minerals industry exercise a disproportionate influence upon the pattern of supply and demand. Such factors reflect the policy priorities of government, for example, by reducing the level of public sector borrowing the provision of public authority housing has been diminished with a consequent fall in demand for construction materials. Policies, programmes and plans, in both the public and private sectors, exercise a direct influence upon the supply and demand for minerals. In addition the strategic policies pursued by national, regional and local government exercise a high degree of influence upon the level and rate of mineral extraction.

The characteristics considered above constrain and condition any procedures which may be used in the forecasting of supply and demand. Any failure to fully incorporate the range of variables which control and influence the generation of a forecast can result in a high degree of variation between the expected or target rate of production and use, and the outturn. Likewise, an inaccurate evaluation of, for example, viability can result in initial errors being compounded. The experience of attempting to predict the level of future demand for sand and gravel in the period after 1945 illustrates the level of divergence that can occur between target and outturn, if an initial estimate of demand, either fails to incorporate the necessary range of variables, or is unable to predict to a satisfactory degree of accuracy future trends in the major controlling factors. The Advisory Committee on Sand and Gravel (Waters, 1948 to 1955) underestimated the demand for sand and gravel, due to the adoption of the incorrect assumption that the overall level of demand for aggregates would fall following the immediate period of postwar reconstruction. In 1952 the Committee's initial estimate of the annual demand for sand and gravel, up to the year 2000, (of 28 million cubic metres per annum) was exceeded. Subsequent estimates of demand for aggregates have taken, over a shorter time period, a far more cautious and detailed view of the difficulties involved in long term forecasting.

DEMAND
Although it is inadvisable to consider demand in isolation from supply it is evident that previously this has frequently occurred. Many forecasting exercises have distinguished the future expected level of demand, (which in reality forms the basis from which supply estimates can be judged) from an analysis of the potential pattern of supply. Recent developments in forecasting techniques have attempted to inject a higher degree of integration into the generation of estimates and have attempted to more closely match supply with demand. An example of this change in approach can be seen in the move away from the use of

a simple energy curve as a basis for estimating energy demand. Until recently it was assumed that the level of demand for energy could be estimated through an examination of forecast rates of gross domestic product formation ; gross domestic product was seen in this case to indicate the potential level of demand to which supply would then naturally react. In the post 1973 period the simple linear energy curve failed to provide a valid explanation of the relationship between supply and demand. It is now recognised that supply does not automatically follow demand and that the price mechanism cannot provide (in isolation) a tool for the regulation of demand (Hutber,1980).

The Verney Committee (1976) outlined the major methods which are available for the forecasting of demand for aggregates :

   i.   by the extrapolation of past production trends

   ii.   by the establishment of statistical relationships between the level of construction activity and the consumption of aggregates, from this a forecast of demand can be based on forecasts of construction sector growth, stability or decline

   iii.   by isolating the demand for aggregates for each type of construction activity and thus basing future demand upon assumptions of growth in the various sectors of the construction industry

   iv.   by relating aggregates demand to particular projects and programmes, for example, by assuming a linear relationship between past and future aggregates consumption in the construction of new roads.

Demand forecasting for the majority of mineral resources has tended to utilise one or more of the methods outlined above. Specific variables are normally incorporated in the forecasting technique, for example, in forecasting the demand for coal the future fuel needs of the electricity production industry and the iron and steel industry are taken into account.

The majority of demand forecasts relate to national or international markets ; assumptions are therefore incorporated into forecasts which imply a uniformity in the pattern of intranational demand. An exception to the general trend is the recent attempt in the United Kingdom to produce, at a regional level, detailed assessments of the likely level of demand for aggregates. In 1969 seven Sand and Gravel Working Parties were established in South East England, these Working Parties were constituted under local authority chairmanship and were based on the gravel regions previously suggested by the Waters Committee. The reports produced by individual working parties were collated by the Standing Conference on London and South East Regional Planning (1974). An Interdepartmental Steering Committee of central government officials was instructed, in 1970, to undertake a preliminary review of aggregates policy. This committee made a series of recommendations, which included the need to establish working parties in all United Kingdom regions ; these working parties were given responsibility for analysing the current supply and demand for aggregates and producing forecasts of future regional trends. By 1975 Regional Aggregates Working Parties had been established in all of the economic planning regions of England, Wales and Scotland. They are linked, through their organisational structure, to the regional conferences of planning authorities, and as such detailed assessments of demand for aggregates, and the availability of resources can be undertaken within the context of regional and county strategic planning.

The demand forecasts produced by the Regional Aggregates Working Parties are based upon the national forecasting model developed by the Department of the Environment (1978). This model utilises a multiple regression format in order to forecast the demand for aggregates in relation to construction output and road expenditure. In the model :

$$D = a + b1 \, C.O. + b2 \, R.E.$$

where D = the demand for aggregates per annum (in tonnes)
C.O = construction output at constant prices
R.E = expenditure on road construction, improvement and maintenance at constant prices
a = a constant term
b1; b2 = coefficients.

An indication of the model's performance between 1972 and 1976 is shown in table 3·2, where the estimated actual level of production of aggregates is compared with the estimate provided by the model.

Table 3·2

Aggregates demand in Great Britain.

Year	Estimated Production	Estimate from the Model
1972	260	264
1973	284	281
1974	257	253
1975	248	241
1976	225	228

Million tonnes
Source: Department of the Environment (1978)

It can be seen that the success of the model is dependent upon the accuracy of other forecasts , this assumption in the current economic climate is questionable. A further doubtful element in the model is the implicit assumption that forecast expenditure targets will be achieved. On the basis of the most recent evidence available regarding motorway and trunk road construction expenditure a 14·8 per cent shortfall in planned expenditure occurred during the financial year 1978 to 1979. (Imber, 1980) An additional factor, which can cause difficulties in forecasting the demand for aggregates, is the weather ; the forecasts make the assumption that the weather each year will correspond with the average for the recent past. During the first quarter of 1978 the weather was poor and production was correspondingly low. (Department of the Environment, 1978).

The current forecasting model incorporates an allowance for uncertainty regarding future production by fixing the most likely estimate of demand within a range. As can be seen, from table 3·3, the range of thresholds increases the further that the forecast moves into the future. For 1980 onwards the upper and

Table 3·3
Forecast of demand for Aggregates, Great Britain, 1977–87.

Year	Forecast Demand	Range
1977	206	198 - 214
1978	213	200 - 226
1979	218	201 - 235
1980	220	198 - 242
1981	223	201 - 245
1982	226	203 - 249
1983	228	205 - 251
1984	231	208 - 254
1985	234	211 - 257
1986	238	214 - 262
1987	241	217 - 265

Million tonnes.
Source : Department of the Environment (1978)

lower thresholds are fixed at 10 per cent above and below the forecast likely level of demand.   Although the adoption of this technique, which utilises a range of estimates, provides the advantage of allowing for a more robust planning response at national level, difficulties are encountered when assigning estimates of demand to the regional level.

Whereas, it is a straightforward exercise to define an estimated national level of demand, the apportionment of such demand to individual regions is far more difficult.   Significant variations exist between regions in the level of construction industry output and expenditure upon road construction.   In addition, it is also difficult to specify, for a particular region, the precise level of demand due to the incomplete coverage of data for the distribution of aggregates (Department of the Environment, 1978).   Shane (1978) has suggested two possible approaches to the resolution of these difficulties. Firstly, indices calculated from base year data could be used to apportion demand to regions.   Secondly, the national model could itself be directly applied at the regional level.   The first of these alternatives has the inherent weakness of assuming constancy in the proportion of demand for each end use in each region. The second alternative would require data which are not fully available for all regions.

Given the difficulties which are encountered in attempting to directly utilise the national model to determine regional demand, and recognising the problems which would exist in adopting Shane's alternative approaches, then the current regional demand forecasts represent a compromise.   Regional demand forecasts are derived from the national model, subject to modification on the basis of available regional data.   Further adjustments are then made by reference to the regional forecasts of the key variables which underpin the national model. From 1978 onwards the regional demand forecasts have incorporated an estimate of regionally relevant motorway and trunk road expenditure and a forecast of local road expenditure based upon the bids made in the county Transport Policy and Programme documents.   In addition, a 1977 regional split of construction expenditure has been applied to the national forecasts of construction output.

Regional forecasts produced by this process are volatile, due mainly to variations in the expenditure programmes of both the public and private sectors. However, the forecasts do provide a framework for the work of individual Regional Aggregates Working Parties.   Individual working parties, despite having reservations about the nationally generated forecasts, have utilised the regional estimates produced by central government, but have also incorporated their own modifications.   The West Midlands Aggregates Working Party in their Regional Commentary (1980) illustrate the style of modification which can be made.   As is apparent, from table 3·4, the West Midlands authorities regard the national regional forecast as too optimistic, therefore, they suggest a modification to the cumulative level of demand.   This modification has the result of reducing the nationally generated middle range estimate of cumulative demand for the period 1978 to 1986 from 172 million tonnes to a lower regional estimate of 150 million tonnes.   An adjustment of this magnitude is significant in that it is based upon the assumption that demand can be met from production within the region.

Table 3·4
West Midlands Cumulative Aggregates Demand, 1978 - 1986

West Midlands Estimate		Department of Environment Estimate	
		top of range	197
top of range	172	middle of range	172
middle of range	150	bottom of range	147
bottom of range	130		

Million tonnes
Source :   West Midlands Aggregates Working Party (1980)

One of the dangers inherent in the adoption of a regionally developed set of estimates is that if, for example, demand should exceed the lower estimate then, either imports of aggregates will have to occur, or production will increase within the region.   If all regions produced their own estimates, (which may prove to be realistic) then the problem of producing an integrated national pattern of demand and supply will be increased.   One solution would be to utilise regional modifications to the key variables, which form the base of the national model, during the early stages of forecasting.   This would allow regional variations to be accounted for whilst retaining overall consistency.

A further difficulty, which again can be illustrated by reference to the West Midlands region, exists in attempting to reconcile short and medium term trends in demand with longer term forecasts.   Although the West Midlands Aggregates Working Party is confident in estimating an annual level of demand, some 3 to 4 million tonnes below the national esti mate, during the period 1978 to 1986, after 1986 the national estimate is assumed to be correct.   The suggested regional modification to the national forecast is justified on the basis of detailed local knowledge.   The question which must be asked is, why was such information not used by the Department of the Environment in the generation of the national and regional forecasts of demand?

Whatever weaknesses may exist in the current attempts to estimate the future regional level of demand for aggregates the estimates which have been produced represent a significant improvement upon the previous situation where little or no information was available.   The generation of national and regional estimates allows planning authorities to conform with one of the main recommendations of the Verney Committee (1976) which suggested that,
'planning authorities in areas of high demand should reappraise the priority accorded to aggregates production with a view to ensuring

continuation of locally produced supplies at recent levels for the
next 10 to 15 years'.

Forecasts of demand for other mineral resources do exist in the United
Kingdom, however, they are frequently produced at national or local levels.
The regional forecasts produced by organisations, such as the National Coal
Board, utilise spatial units which are not coincident with county or economic
planning region areas.   Some of these estimates suffer from the inherent
weaknesses which have already been illustrated, for example, they assume a
constant relationship between the level of economic activity and resource
consumption and anticipate a constant pattern of demand.   These failings
result in a lack of flexibility in provision to meet new forms of demand, for
example, it is claimed that new industrial development in certain regions is
inhibited due to the lack of a readily available supply of gas.

From the preceeding paragraphs it can be seen that many of the difficulties
that are encountered in understanding the current pattern of demand, and
therefore in forecasting future demand, stem from a lack of suitable data.
Forecasts are alas subject to uncertainty because sufficiently reliable sectoral
consumption estimates are not available.   At a regional level demand levels,
(at present and for the future) are only available for a limited range of mineral
resources.   Inevitably, the estimation of demand is linked to an understanding
of the pattern of production and distribution ; the analysis of supply provides
the focus for attention in the following section.

SUPPLY

The current production of mineral resources in the United Kingdom is well
documented.   Annual production figures for a wide range of minerals are
available at county level; in addition, data on minerals production by end use
can also be obtained.   (Institute of Geological Sciences, 1980).   A far more
difficult matter is the estimation of reserves, which are or will be available, and
therefore the future level of supply.   Quite clearly (although not always
apparent at the regional scale) supply at one point in time is related to the level
of demand, and is thus subject to fluctuations in the level of economic activity.
The responsive nature of minerals production can be seen in situations where
increases occur in order to satisfy a short term demand, as in the development of
borrow pits to provide materials for new road construction.   Alternatively, as in
the cessation of iron ore mining consequent to the closure of iron and steel
production facilities, reductions in output can occur.

Minerals production is constrained and controlled by a number of limiting
factors; firstly, by the extent to which a mineral can be regarded as ubiquitous;
secondly, by the costs of extraction, processing and transportation; thirdly, by
the availability of alternative or substitute materials; and finally, by the nature

of public and private sector controls upon production.

Few minerals in the United Kingdom can be regarded as truly ubiquitous; Blunden (1975) has identified certain aggregates (sand, gravel and some hard rocks) and other non metalliferous minerals (chalk, limestone, clay, shale and brick clay) as being relatively ubiquitous. These minerals can be distinguished from localised minerals such as potash, fuller's earth, china clay and the majority of metalliferous, carbon and hydrocarbon minerals. The distribution of available minerals resources gives rise to a pattern of production and transportation that reflects the complex interaction of market forces and the pattern of control which is exerted upon the extraction of minerals.

The costs involved in mining operations vary through time and between locations. As a general rule, (although there are notable exceptions) mining activities in the United Kingdom have tended to become increasingly concentrated in larger workings, under the control of larger enterprises. This tendency reflects both the increasingly high capital and operational costs of extraction, and the desire of operators to rationalise their processing and transport arrangements. Significant economies of scale can be obtained through increasing the level of output from a single location; larger operations also allow for the risks involved in new developments to be shared between a number of operators. This tendency to concentrate production at a smaller number of workings can be seen in the changing pattern of coal production in the North Yorkshire area of the National Coal Board. Coal mining in this area has become increasingly concentrated on a smaller number of collieries, (from 1963 to 1977 the number of collieries was reduced from 30 to 18) with some loss of output, (from 10,400,000 tons to 8,100,000 tons) but with a significant increase in output per manshift (from 38·2 cwt. to 50·6 cwt) (North and Spooner, 1978). In a similar manner sand and gravel operations have increased in size, not least in response to the changing nature of transport costs which have encouraged the use of larger road vehicles and, where available, rail haulage. Cost sharing, in order to reduce individual risk, can be demonstrated by the frequent adoption of a joint development and trading mode of operation by companies involved in the extraction of offshore hydrocarbons.

Competition from alternative and substitute sources of materials also affects and alters the pattern of mineral production. The proportion of United Kingdom sand and gravel obtained from marine dredging has increased from 5·8 per cent of total production in 1958, to 10·1 per cent in 1968 and to 14·3 per cent in 1978. This illustration demonstrates the changing pattern of supply for one type of ubiquitous mineral which is also, in part, being replaced by substitute materials. In 1977 the supply of alternative materials was equivalent to 7·6 per cent of the total production of aggregates in England and Wales. (Regional Aggregates Working Parties, 1980). The use of substitute and recycled materials tends to increase in direct relation to commodity prices, and to the possibility of utilising low cost materials which were previously regarded as waste products.

Public and private sector decisions on matters of policy and operation constrain and control the supply of mineral resources. At one extreme, central government policy regarding the rate of extraction of offshore hydrocarbons exerts a direct influence upon the quantity and price of United Kingdom oil and gas supplies; at the other extreme a local planning authority decision, as to what constitutes a material change of use, can directly affect an individual mining operation. Private sector organisations also have a direct influence upon the supply of minerals; the adoption of a specified rate of return on capital invested in a given project can jeopardise or eliminate a source of supply of a mineral which is subject to market price fluctuation.

Given the above constraints upon the production of minerals it is, in theory, possible to base estimates of reserves and future production upon anticipated demand. In practice there are two main obstacles to such a simple calculation. Firstly, in examining, for example, an individual oil field it is possible to calculate the annual yield and life span of that field. However, the proportion of a known oil reserve which is extracted will depend upon prevailing market conditions, and upon the cost of extraction which tends to increase as the size of the reserve diminishes. Secondly, it is difficult to base estimates of future production on forecasts of demand, due to the uncertainty which surrounds the demand forecasts. For the majority of mineral resources a distinction can be made between, the level of production which is technically possible, and that which is commercially probable.

Certain minerals only occur in a limited number of regions. In such cases it is unnecessary to undertake an interregional evaluation of production, although the regional pattern of consumption is of fundamental concern to the planner. Aggregate minerals are relatively ubiquitous within the United Kingdom, and since the mid 1970's production and reserve figures have been available, in a systematic and co-ordinated form, for regions and for county areas. As can be seen from table 3·5 there are significant intraregional discrepancies between the production of aggregates and the level of demand. These differences create a situation whereby imports and exports of aggregates occur; certain regions are deficient in particular forms of aggregate, while other regions are net exporters of all aggregates. Six of the ten aggregates regions are net exporters, the other regions import large quantities of certain types of aggregate. The total production and consumption data does not however reveal the complexity of supply patterns. In all regions the total consumption of aggregates exceeds the level of consumption from production within the region, the East Midlands region, for example, was in 1977 a net exporter, but also imported 1,288,000 tonnes of aggregates from other regions.

35

Table 3·5
Production and Consumption of Aggregates – England and Wales 1977

Region.	Aggregate Consumption.	Consumption produced from production within Region.	Aggregate Production.
South East.	46,647	38,463	39,733.
East Anglia.	8,876	7,861	8,609
East Midlands.	17,428	16,140	26,990
West Midlands.	17,431	15,305	17,980
South West.	19,867	18,481	25,499
North West.	17,280	8,430	8,713
Yorkshire.	16,571	12,804	15,057
Northern.	14,326	12,675	14,441
South Wales.	11,511	11,232	12,100
North Wales.	3,487	3,298	5,970

Thousand tonnes.
Source: Regional Aggregates Working Parties, (1980).

   The transfer of aggregates between regions indicates the level of the supply
shortfalls and production surpluses which occur.   In certain regions there has
been a total production shortfall.   In 1979 the West Midlands region consumed
3·8 million tonnes of crushed limestone, of which 1·3 million tonnes were
imported from the East Midlands, South West and North Wales regions.   By way
of contrast, the West Midlands consumed 4·4 million tonnes of crushed igneous
rock and sandstone, of which 0·1 million tonnes were imported even though
regional production of these aggregates totalled 5·1 million tonnes.   Such
seemingly inefficient export and import patterns are, in practice, inevitable
given the particular characteristics of the minerals which are produced, the
structure and ownership of the minerals industry and the form of the transport
network.   The North West Region, for example, is a net exporter of sand,
although in three of the four constituant county areas there is a shortage of
concreting sand. (North West Aggregates Working Party, 1979).   Ownership
patterns are reflected in a desire of some operators to use their own sources of
supply wherever possible.   Transport availability has a marked effect upon the
interregional movement of aggregates.   As Verney (1976) commented 'transport
is the key to many of the problems surrounding the supply of aggregates'.   The

use of water or rail transport can significantly increase the distribution area from a point of production, due to the lower tonne mile costs which are incurred by the consumer. The movement of aggregates is therefore indicative of scarcity and price differentials as well as the natural endowment of regions.

A further major factor which must be considered when examining the regional pattern of aggregates supply is the location and size of reserves. For the purpose of forecasting, the amount of permitted reserves is of great significance in that it indicates the likely future level of supply of a range of aggregates. In recent years county and regional authorities have adopted the general principle suggested by Verney, (1976) that reserves should be equal to ten years supply at the current level of consumption. However, in many regions this target for permitted reserves has not been achieved. As a consequence a topping up procedure is used, whereby land is granted planning permission at a scale which is sufficient to replace that which has been worked out (Bate,1980). Without a precise definition of the level of permitted reserves, and in the absence of overall direct control over the rate of extraction, then serious shortfalls in production are likely to occur in the future.

The supply estimates, which are now available for aggregates, indicate the complex nature of the regional distribution and production of mineral resources. It is necessary to understand the complexity of the situation before realistic strategic planning can take place. A parallel detailed regional dimension is required for other minerals. This was recognised by the Stevens Committee (1976) who identified the need for, 'advice on all these minerals, (that is, coal, iron ore, fireclay, clay and shale, slate, limestone, sandstone, sand and gravel, igneous rock, gypsum, anhydrite and fluorspar) not just on those for which regional working parties are established'. There is a strong economic argument, if future mineral resource development is to be effectively managed, for the extension of the work of the Regional Aggregates Working Parties to encompass other minerals of strategic significance.

THE REGIONAL DIMENSION

The development of mineral resources in the past and at present has frequently provided the impetus for regional growth. Certain minerals such as coal, have a direct effect upon regional development, whilst others have a less obvious relationship with their region of occurance. The importance of the relationship between minerals and national and regional economic growth was recognised by the Stevens Committee (1976), who claimed that, 'it is difficult to exaggerate the importance of minerals to our way of life, our industry and our economy'. This concern for the relationship that exists between the occurance and extraction of minerals and matters of central concern to regional and strategic planning is not confined to economic and industrial issues. Of equal importance are the social, land use, environmental and spatial consequences of mineral resource development.

During the past decade the development of North Sea hydrocarbons has exerted a significant influence upon the strategic planning of North East Scotland. With the discovery of commercially viable reserves of oil, many of the planning policies of local, regional and national authorities have been subject to adjustment in order to ensure that the needs of the offshore industry were met. The need for the onshore provision of industrial sites and infrastructure has necessitated amendments to previously accepted plans and strategies. An early example was the amendment to the Ross and Cromarty Development Plan, whilst, more recently, the Structure Plan prepared by the Grampian Regional Council reflects the requirements of the offshore industry (Roberts and Shaw 1977).

Most of the activities which are associated with the offshore industry make demands upon the infrastructure of the area within which they are located. In many instances activities were developed in hitherto rural areas, where the infrastructure base was insufficient to meet the new demands which were generated. The establishment of rig and platform construction yards in rural areas required the building of new roads, the provision of additional housing and a wide range of local services. Major urban centres, such as Aberdeen, were also required to rapidly extend their infrastructure base. Aberdeen exhibits many of the consequences of offshore associated development; major port and airport improvement has occurred, additional industrial land has been designated, the housing stock has been increased and road and rail links to the rest of the United Kingdom have been improved (Lewis and McNicoll, 1978).

North East Scotland has experienced a number of impacts related to the development of the offshore industry; the creation of major new industrial facilities, the establishment of service bases, the construction of reception terminals and the indirect and induced effects of increased demand. In each case these impacts have had economic, social, land use, environmental and spatial consequences. The planning response to offshore related developments has occurred at national, regional and local levels. At national level there has been an attempt to encourage the offshore industry to locate its activities in 'preferred development zones', and to restrict the growth of new development in 'preferred conservation zones' (Scottish Development Department, 1974). Planning guidelines have also been prepared in order to assist regional and local authorities to accommodate the needs of the offshore industry. One response by regional authorities has been to encourage new developments to concentrate in key settlements. In Easter Ross this has been achieved by the expansion of housing in Invergordon, Alness and Evanton, and through the provision of industrial land adjacent to these settlements and along the Cromarty Firth.

As well as stimulating the growth of new and expanded settlements, and requiring the provision of new and improved infrastructure, the offshore industry has had a profound effect upon the structure and performance of the Scottish economy. Although estimates vary, it is calculated that, by 1976, some 38,000 oil related jobs existed in Scotland. In the Grampian Region oil related employment, in 1976, represented 7·3 per cent of total regional employment, whilst in the Highlands Region 9·3 per cent of total regional employment was in oil associated activities. The multiplier consequences have also proved to be significant, by 1976 oil related employment was estimated to be between 55,600 and 64,800. (Lewis and McNicoll, 1978).

From this brief review of the effects of the extraction of North Sea hydrocarbons upon North East Scotland, it can be appreciated that the regional dimension of resource development is extremely significant. Although the example which has been presented illustrates the range of issues which confront planners in somewhat exaggerated terms, nevertheless, the principles involved are capable of transfer to other situations. It is clear that a regional economy can become inextricably linked to the extraction of a resource, and that a reduction in either the supply of or the demand for that resource can undermine the economic base of the region. Many parallels can be drawn between the development of offshore hydrocarbons and the effects upon Scotland and the extraction of other minerals. Blunden (1977) comments that 'stock resources as a basis for economic development can, therefore, present long term problems, perhaps made all the more pertinent where they are the mainspring of primary and secondary employment in the area.' Whilst it is apparent that minerals play a significant role in the growth and decline of regions, it is also clear that the transitory nature of mineral extraction presents planners with a number of major problems. In order to satisfy the demand for minerals, and the demands of the minerals industry a planned strategic solution is required which incorporates estimates of the future pattern of supply and demand, identifies the nature and scale of future difficulties and, if possible, resolves those difficulties before they become severe.

4 Needs

RENEWABLE AND NON RENEWABLE RESOURCES

The approach that governments and societies in different regions of the world have adopted for the utilisation of their natural resources varies considerably. Economic, ecological, social and broader strategic factors all exert influences upon the way in which nations perceive and respond to natural resource exploitation and management. Minerals must be considered as a part of the natural resource heritage of all countries and regions. As national economies develop, their demand for minerals from domestic and foreign sources increases. Although the requirements for certain types of mineral has declined, others are in increasing demand and this is likely to continue with the emergence of third world nations.

There has been, until recently, a tendency throughout the developed world to think of most essential minerals as existing in abundance, perhaps with the exception of precious metals and gems. Although European countries experienced significant problems of mineral supply during the two world wars this was viewed as a temporary situation which would be rectified at the termination of hostilities when the normal pattern of supply was restored. Scarcity of mineral resources is a relatively new concept and has prompted a wide range of groups in society to review both national and international situations in respect of the mineral requirements of developed and emerging nations. Such debates have become more common during the past two decades and the distinction between renewable and non-renewable resources has become more commonly accepted.

Warren and Goldsmith (1974) have developed useful definitions of these two categories of resources based on the earlier work of Ciriacy-Wantrup (1952). They make a simple distinction between renewable resources which they term 'flow' and non-renewable resources which they call 'stock'. Their definitions are for (i) non-renewable resources (stock),

'materials that are concentrated or created at rates that are very much slower than their rate of consumption',

and for (ii) renewable resources (flow),

'parts of functioning natural systems that are turning over at rates which are approximately comparable to their rate of use'.

Governments and industry have been forced to review their position for current and future minerals requirements not only through pressure from ecologists and environmental groups, but also through their own fear of running out of natural resources. Scarcity is not simply a question of using up all available sources of a mineral or so mismanaging a renewable resource that it takes on a finite dimension, but it can be induced by other factors. Blockades, the refusal of one nation to supply another at all, or at an acceptable price, can also induce a shortfall in supply. Although the international arena highlights this situation through an example such as oil trade with an array of social, economic and political implications, scarcity with all its attendant problems can occur in a domestic situation, for example, planning policies in the United Kingdom National Parks have had an impact on the flow of limestone to industry, and therefore the economic performance of construction activities.

In any consideration of mineral resources, which are generally finite in nature, the time period over which they are to be exploited may well be critical. It is invariably a tenet of good management that any resources should be husbanded over the longest period of time for the greater good of humanity (McVean and Lockie, 1969). Where local or national economies depend upon the working and free flow of mineral resources such a viewpoint is not always acceptable, either to government and industry for whom profit maximisation, strategic considerations and political expediency may be more important, or to those who live in communities which depend upon the winning, transport and utilisation of mineral resources for their livelihood. It would therefore be naive to think that in either a national or international context governments were likely to support or adopt a purist approach to conservation which unduly restricts the supply of mineral resources.

However, examples do exist of governments attempting to develop a rational conservation policy for particular minerals. During his administration President Carter attempted to introduce a conservation of energy policy in the United States. Clearly much of this policy revolved around oil. Domestic, commercial and industrial consumption of oil in America is one of the highest in the world, approximately 36 per cent of total world consumption (Dasmann, 1976). Many of Carter's proposals were directed at the very high rate of oil consumption for domestic purposes, petrol, central heating and air conditioning. This conservation policy was to be introduced by the restriction of supply, an increase in price through the removal of the Federal Government subsidy on imported crude oil and also through a campaign of public enlightenment. Carter's energy strategy met with little success and in fact generated considerable hostility from industry and the general public.

In contemplating the introduction of conservation measures which suggest at the very least a slowing down in the rate at which mineral wealth is consumed governments are forced into the realisation that, for many mineral commodities, demand is largely inelastic and that societies which have grown wealthy upon the exploitation of the world's natural resources find it particularly difficult to countenance policies which are likely to disrupt life style and personal expectation. Countries which are large consumers of mineral wealth also face the short term hazards of adding to inflation by restricting the availability of basic raw materials. For many governments then, there is the problem of balancing longer term strategic issues with short term economic problems.

The easiest approach for governments to take when faced with such a dilema is to look for alternative ways of coping with strategic issues for the future whilst allowing existing mineral resources to be exploited at the same or increased rates. This has been referred to as a 'brinkmanship' approach where future needs and security are dependent upon the ability to discover new supplies of the same resource or to discover alternatives. Government interest in considering alternatives is variable. For example, in the field of energy given the finite nature or carbon and hydrocarbon resources most western countries have invested heavily in nuclear power generation projects but have partially ignored the arguably more environmentally acceptable alternatives for power generation, notably solar, wind and wave. There is a fear in many quarters that the refusal by national governments to undertake a major review of natural resources and approaches to their consumption now, will leave nations with insufficient reserves for the future.

Although much of the concern about mineral resources is centred around a too rapid depletion of finite stocks, concern is also expressed about the associated implications of mineral winning on the environment. Ecological degradation has been caused through the extraction and subsequent use of mineral resources, and although in some countries and regions governments have made an effort to minimise some of the more harmful effects of minerals winning and their after use (Olschowy,1971), inevitably economic and political pressures will dictate a continuation of these activities.

It would be unrealistic to think that there will be any radical alteration in the organisation of industrial development in any part of the world in respect of fundamental requirements of raw materials. What however might be possible is that governments respond to the words of ecologists and environmental economists and look for alternative ways of satisfying the needs of industry, commerce and the domestic consumer.

ALTERNATIVES AND RECYCLING

The search for alternatives to essential minerals is something which has long exercised the minds of environmentalists, economists and governments. Indeed, during both world wars this was vital and various innovations were noted but unfortunately in all too many cases such ideas were quickly shelved and traditional resources reinstated. A number of alternatives to petrol, for instance, were in use during the second world war but until recently few continued to be used as commercial alternatives. Although some groups press for and sponsor research into alternatives on environmental grounds, most of the thrust for alternatives has been brought about by economic necessity and through strategic considerations.

Recycling is arguably the most common example of the way in which demand for raw materials can be met without total recourse to virgin stocks of minerals.

Both economic and environmental factors motivate recycling though there is a presumption that appropriate technology is available. The reuse of scrap metal was, for example, dependent upon the development of technology that was capable of reusing a wide range of ferrous products of varying quality. It is unlikely that scrap would ever have become a feed stock had not its price compared favourably with that of iron ore and the high cost of manufacturing pig iron, but during recent years the high energy costs or resource extraction and primary processing have brought this practice closer to the margins of economic viability.

Colliery spoil heaps have been used, where the waste material was suitable, as a basis for road construction, for example, in part of the construction of the M.I Motorway in South and West Yorkshire. This represented a saving of newly extracted aggregate material. Despite the availability of large quantities of such waste in different parts of the country one major factor has hindered the extension of this practice, namely the high cost of moving bulk, low value goods over long distances. This question of costings will be developed at a later stage in this text, suffice it to say here that in the context of a broader approach to the evaluation of projects, what has hitherto been a useful and a cost effective way of using waste material as a bulk fill in preference to newly worked minerals at a local level, might well have a more extensive application regionally and nationally.

There are many factors which can influence the availability of waste products for reuse even when they are present in the immediate vicinity of development works. For example, the use of unburnt colliery shale in road construction has long been deemed injudicious for a number of reasons, not least of which being the danger of spontaneous combustion (Oxenham,1966). Given the exhaustion of stocks of red, burnout shales in many areas, there has been increasing pressure during the past decade to permit the use of unburnt shales for certain categories of construction work. (Road Research Laboratory, 1968).

The by-products of mineral winning operations can have quite specific uses. Prior to large scale mechanisation below ground in coal mining, much of the coal and other materials were hand sorted. 'Blue Stone', a material occurring in certain coal measure series, was particularly suited for brick making and in the period before nationalisation and again after 1946, private coal companies and then the National Coal Board had quite profitable brick making works close to their coal mines. Mechanisation below ground meant that the material was not as well sorted and the additional costs of separating out the 'Blue Stone' brought the economic viability of these brick works into question, this means, taking new clay workings elsewhere. However, in the opencast process the various by-products are more easily separated and once again the 'Blue Stone' can be more profitably used in brick making. (Lambert,1981).

Large extracting companies have found it profitable to invest in a dredging capability to exploit off-shore sand and gravel reserves, partly because of the depletion of existing workings and partly due to mounting opposition to new and extended workings. Despite an increase in the amount of aggregates derived from marine won sand and gravels in the post war period such operations have not been without their critics. Concern has been expressed about potential damage to marine habitats and to coastal areas through erosion (Nature Conservancy Council,1979). Similarly, foreshore operations have come under greater scrutiny in recent years because of the problems they pose to coastal erosion (Merseyside County Council 1979). Despite advice available from bodies such as the Hydraulic Research Station, it is likely that amenity and ecological groups will continue to harry both the Crown Estate Commissioners (who grant licences for off-shore operations) and local planning authorities (who grant planning permission for foreshore operations) to reduce sand and gravel working in these areas.

Future investment in the search for alternatives to existing minerals and alternative sources of traditional materials will be conditioned by a number of factors. Falling demand for a particular mineral is one such factor. In the United Kingdom although it is still possible to identify regional shortfalls in demand over production, there has been a steady decline in the demand for aggregate minerals during the past five years. Government forecasts suggest that by the late 1980's and the early 1990's there will once again be a high demand for aggregate minerals (Department of the Environment,1978). It has been noted however in a recent report from the Standing Conference on London and the South East Regional Planning (1979) that even if resources have been identified and licences from government are forthcoming, 'the level of confidence in the industry (and) the availability of capital investment' are likely to have a dramatic impact on future operations. Without adequate thought being given to the future by operators and the government it is not inconceivable that there will be an inability to satisfy demand. It takes three years to build a new dredger, it takes time to identify new resources, and money and time are required to invest in appropriate technology for future works. The current attitudes of operators and governments make substantial investment in alternative strategies for the future unlikely. Longer term planning must be seen as desirable if future demand is to be satisfied and a comprehensive strategy devised for the utilisation of mineral wealth which takes account of traditional minerals as well as alternative mineral resources.

There has been in operation since 1966, set up under what was then the Ministry of Public Buildings and Works, an Agrement Board. The function of this Board is essentially to assess and certificate new construction materials and systems. Clearly it would be unwise to utilise waste products as alternatives to natural mineral resources in construction if their suitability and safety had not been tested.   Although it is true that industry invests in the search for new and alternative products,a case may well exist, if the most effective long term use of minerals is to be achieved, for central government to direct additional money into research.

Reaction to alternative sources of minerals has generally been a pragmatic response to the fluctuations of supply, demand and cost.   There is a case to be argued in favour of producing a co-ordinated policy on minerals supply which takes into account the full range of sources and of waste, recycled and alternative products.   The feasibility of doing this at a regional or national scale will be considered at a later state in this text, but it is certainly worth noting that in both the report of the Stevens Committee (1976) and the report of the Verney Committee (1976) this approach was advocated.

ECONOMIC AND SOCIAL ISSUES

All planning decisions regarding mining activities, either incorporate the economic and social costs associated with an activity, or by implication apportion such costs.   Broadly, a distinction can be made between the national, regional and local incidence of costs and benefits.   Arguments as to what constitutes the interests held by the various parties associated with mining activities are discussed in Chapter six of this text.   The link between the need for mineral resource development and the economic and social characteristics of regions provides a focus for the present discussion.

Economic issues are of major significance in considering the need for new or existing mining activities.   Not least of these issues is the continuing role played by mineral extraction in the provision of employment.   In a number of British regions employment in mining is substantial, and whilst the level of total employment may change through time the legacy of previous patterns of economic activity remain.   Table 4·1 shows that there is a concentration of mining employment in four regions, these are areas which have a significant coal sector when compared to the national average.   A strong relationship exists between the pattern of economic activity stimulated by the development of coal reserves and the overall performance of a regional economy.   The multiplier effect of coal mining activities is significant, Spooner (1981) estimated a modest multiplier of 1·5.   The majority of mining activities are transitory, however, the activities dependent upon mining and the settlement pattern which results from mining are far less ephemeral.

In many of the United Kingdom's peripheral regions past and present mining activities have left their mark on host areas.   Slate quarrying in the North Wales has, since the late eighteenth century, exerted a strong influence upon the region's economy, social structure and pattern of settlement.   Davies (1974) traced the rise of the industry and showed the rapid creation of an industrial labour force in a hitherto agricultural area.   In one quarry alone (the Penrhyn quarry at Bethesda), the workforce grew from 300 employees in 1802, to 1,000 in 1816, and by 1840 2,000 were employed.   The growth of the labour force was accompanied by urban development; the population of Bethesda grew from 2,600 in 1801 to 8,000 in 1841.   An industry which in 1800 employed a few hundred by

Table 4·1
Regional Employment in Mining, 1976

Region.	Number Employed*	Proportion of Total Regional Employment**
South East.	11·8	0·16
East Anglia.	2·5	0·37
South West.	11·2	0·74
West Midland.	26·0	1·19
East Midland.	71·2	4·76
Yorkshire and Humberside.	82·1	4·17
North West.	14·7	0·56
North.	49·6	3·95
Wales.	41·2	4·14
Scotland.	35·3	1·70
Great Britain.	345·6	1·57

* in thousands   ** as a percentage of total employment.
Source: Department of Employment (1977)

1898 provided approximately 16,700 jobs, mainly in Caernarvonshire and Merionethshire.   From 1898 the numbers employed in slate quarrying feil, by 1935 to 7,6200 and by 1970 it was clear that 'quarrying is no longer a major employer in rural Wales'.   The rise and fall of this mining activity is still visably imprinted upon the social and spatial structure of North Wales.   Attempts at settlement rationalisation have been limited, whilst new jobs have not been created in any number.

A more dramatic, yet similar, history of regional growth and decline can be seen in other regions.   Probably the best documented attempt made by a planning authority to adjust the economic, social and spatial pattern of a mining area is provided by the Durham County Council Development Plan (1951).   This attempt by Durham County Council to consolidate the settlement pattern of a declining coalfield area was notable for its detailed analysis of the locational and physical potential of 357 settlements and the classification of settlement into four categories. Two of these categories, (the C and D villages) were to be stabilised or gradually run down. The objectives of the Development Plan were to concentrate the future distribution of the population in order to make the best use of the limited financial resources which were available. This approach was modified in 1964 (Durham County Council,1964) when certain of the settlements were reclassified, and with the reorganisation of local government some settlements initially scheduled for decline were reassessed (Tyne and Wear County Council,1975).

Although mining areas inevitably experience decline, in Britain it is possible to point to North Scotland, and especially to the Shetlands, as examples of resource induced growth. As has been demonstrated in the previous chapter oil and oil associated developments have had a profound impact on the Scottish economy. The social structure and the infastructure and settlement pattern has also been affected. In a number of cases of oil associated growth there have been strong local reservations regarding the undesirable effects of development. To attempt to resolve such difficulties, the Shetland local authority promoted their own legislation in order to control and gain a share of any development. The 1974 Zetland County Council Act provided the local authority with powers to acquire (through compulsory purchase) land which was required for oil associated development, and to establish a fund from oil revenues in order to ensure that the local community might obtain a direct long term benefit from any development which occurred.

The Shetland initiative is unique in the United Kingdom, the normal method of limiting the undesirable effects of mining activities is through the use of specific conditions which are attached to individual planning permissions. However, such conditions generally relate to the mining operation itself and have little regard for the broader local area or region within which the mining activity is taking place. The present authors have argued elsewhere (Roberts and Shaw 1980) that major mineral resource development should be subject to far wider considerations than simply issues which are confined to the actual point of extraction or processing. If planners do not wish to be faced with the problem, so well described by House and Knight (1967) and Rees (1978), of declining resource based communities, then fuller attention must be given to economic and social issues in deciding upon applications for the extraction of minerals. Set against these local and regional considerations are questions related to the economic need for resources upon which modern society so heavily depends. The need for minerals, as an input into the economy, is discussed below.

NEED AND THE ECONOMY

Mineral resources are fundamental to a modern advanced economy. In their evidence to the Stevens Committee (1976) the National Environment Research Council and the Institute of Geological Sciences observed that; 'It is a human charact eristic to make increasing use of minerals to modify the living environment. A developed society is entirely dependent on the availability of a wide range of minerals. Indeed the degree to which a country is developed could be expressed in terms of its per capita consumption of minerals. The United Kingdom demand for most minerals and mineral based commodities, including metals, continues to increase, for some exponentially'. This observation reflects not only the need for minerals, but also the need to manage resources efficiently; stock resources should be used to maximum effect and flow resources should be managed in order to ensure continuance of supply. As has been already noted there is a constant need to recycle, and to identify substitute sources or forms of mineral material.

Mineral resources are extremely important as an input to the national economy. The Stevens Committee (1976) estimated that, by value, imported minerals (other than coal and aggregates) represented 83 per cent of the total United Kingdom level of supply. Although the situation for some minerals has changed since 1976, for example the import and export balance for petroleum has moved towards self sufficiency, for the majority of non energy and non aggregate minerals the United Kingdom is still dependent upon imports. The level of national self sufficiency which it is possible to achieve is related not only to the availability of indigenous reserves, but also the cost of extracting those reserves or producing alternatives.

The high level and cost of importing minerals prompted central government to take action in order to stimulate a higher level of national self sufficiency. In 1972 the Mineral Exploration and Investment Grants Act introduced an incentive scheme, whereby 35 per cent of the cost of mineral prospecting and the assessment of discovered resources could be reclaimed through a grant in aid. This scheme has resulted in widespread prospecting, mainly for non ferrous metals. The recent exploratory drilling programme, undertaken by Amax Exploration, for tungsten in Devon illustrates the importance of extending indigenous supplies. In this instance the strategic and financial significance of attaining national self sufficiency is clear; in 1977 over £20 million of tungsten and tungsten ore was imported. (Mead,1980).

At national level the importance of minerals can be seen in terms of the interindustry flows that occur in the national input and output tables. Thirty of the thirty four sectors represented in the tables make purchases from the coal mining and other mining and quarrying sectors. Minerals provide major inputs, by value, to a number of sectors; coke ovens, iron and steel, chemicals, bricks, paper and printing, construction, gas and electricity are amongst the major direct purchasers. In addition mineral industries make a major contribution to gross domestic product, in 1979, petroleum and national gas contributed £5,111 million, whilst other mining and quarrying provided £2,696 million (Central Statistical

Office,1980,a).

Mining and quarrying is also a major activity in a number of regions. Although mining and quarrying contributes 1·81 per cent of the total gross domestic product generated in Great Britain, in the East Midlands, Yorkshire and Humberside, the North and the Welsh standard regions over 4 per cent of regional gross domestic product is generated by mining and quarrying. Comparing table 4·1 with table 4·2 it can be seen that there is a high regional coincidence, in the significance of mining activities, between gross domestic product formation and employment provision.

Table 4·2
Gross Domestic Product : Mining and Quarrying.

Region	G.D.P.Formation 1979*	Proportion of total regional G.D.P**
South East	94	0·2
East Anglia	20	0·4
South West	81	0·7
West Midland	202	1·4
East Midland	653	6·2
Yorkshire and Humberside	696	5·2
North West	116	0·6
North	362	4·4
Wales	307	4·4
Scotland	312	2·2
Great Britain	2843	1·8

* £ million   ** as a percentage of regional total.
Source :   Central Statistical Office (1980,b).

At county level the importance of mining is more varied. In the county listings produced by the Institute of Geological Sciences (1979) whilst all counties contain a limited number of mineral resources, (chiefly aggregates) some counties produce a vast range of minerals. Although gross domestic product data are not available at the county level, it is obvious that counties such as South Yorkshire, Nottinghamshire and Durham are heavily dependent upon mining activities. The importance of resource development is reflected in the Structure Plan and other policies adopted by these counties.

Given the importance of mineral resources to both national and regional economies, it is vital that the planning system, whilst accommodating the needs of specific operators and consumers, should take account of future requirements and, more importantly, anticipate shortages in supply. The provision of better geological information and the development of alternative and substitute forms of material provide possible directions for future policy. It is also necessary for the planning system to improve its knowledge of the requirements of industry and to develop more sophisticated methods of relating specific applications to national and regional strategic requirements. This is not to suggest that a family of plans should be developed for minerals from national to local level, rather that explicit policy guidelines should be available to local authorities. The rather negative response of the Department of the Environment (1978) to the Stevens Committee fails to provide adequate guidance. Any response which states that county planning authorities should continue to 'determine each application on its merits having regard to development plans and to other material considerations', where 'other material considerations' are not specified, is unhelpful and could work against long term requirements.

5 Conflicts

One of the prime aims of post war planning legislation has been to avoid or minimise conflict occurring in the process of land use planning. In an effort to provide a coherent context in which decision making can take place policies and strategies have been formulated partly in central government and partly by local planning authorities. The intention of this has always been to ensure that the legitimate development interests of both the public and private sector are not unduly hindered and that at the same time individuals and the public at large are not asked to bear too great, or too disproportionate, a share of the costs of development. Despite attempts to create a sensible framework in which development can take place, and despite the attempts of governments of different political persuations to ensure that some form of equity exists for all those involved, criticisms have all too often been levelled at the decision makers for being too arbitrary in their activities or for bowing too readily to the pressures of short term expediency.

It would be foolish to pretend that planning could always best serve the interests of the public by a rigid adherence to predetermined policies and strategies. Economic, political and technological changes must influence the way in which development proposals are considered and it is clear that if there were no flexibility then opportunities to capitalise upon new development proposals would be lost and this could well be to the detriment of society. It would seem therefore that whilst accepting the need to have a system capable of accepting deviations from previously accepted policies it is of paramount importance that the likely consequences of actions based upon such amended policies are carefully evaluated before any permissions for development are granted. Too rapid and too unconsidered a response to new opportunities can so easily result in diseconomies to both the developer and the public.

As demonstrated in the preceeding Chapter it is unlikely that the demand for minerals will significantly decline in the foreseeable future and even allowing for fluctuations in demand there is every indication that societies throughout the world will continue to have specific needs for minerals. Therefore, the types of conflicts that have arisen as a result of mineral winning operations in the past are likely to remain unless the context for minerals planning can be greatly improved. It would of course be naive to assume that even with improvements to the planning system conflicts of interest would not arise or have to be dealt with. What can be reasonably hoped for in a revised approach to minerals planning is that greater attention can be paid to the true nature of public and private costs and benefits and that this can form the basis for introducing a greater degree of equity into the planning process.

Conflicts arise for a number of reasons, ranging from the threats new minerals developments pose to the environment, to the impact they have upon existing land uses. In some cases the conflicts are generated because of discrepancies in the policies of different government departments. In many cases the appropriate local planning authority will endeavour to resolve any conflicts that arise from a particular application by, for example, imposing conditions upon the developer in respect of after care and mode of operation, or by means of compensation payments.   Those parties who are not satisfied by the decision of the local planning authority may seek to make representation to the Secretary of State for the Environment.   In extreme cases it is possible that a decision made by a Secretary of State might be reversed in Cabinet or as a result of a Parliamentary vote.   It should be appreciated that conflicts do not simply arise as a result of specific applications but also as a result of policies and strategies presented in the form of Structure and Local Plans.   The 1971 Town and Country Planning Act makes provision for these policy documents to be debated and criticisms raised prior to them being ratified by the Secretary of State.

The ways in which planners address themselves to the issue of resolving conflict and the extent to which they attempt to balance public and private costs is developed in Chapter 9, 'Testing, Evaluation and Assessment'.   In this Chapter further consideration is given to the types of conflict which have arisen in the post war period between mineral operators and other groups within society. It is hoped that this will provide a context for a more detailed debate about the improvements that could be made to that part of the planning system that deals with minerals.

URBAN DEVELOPMENT

The past century has been a dramatic increase in the rate of urban growth in the United Kingdom.   In terms of the amount of agricultural land taken annually for urban development the period between the two world wars was the most dramatic.   Although the conversions of agricultural land to urban growth did not attain the same annual peaks in the post war period, nevertheless substantial areas of land were developed.   It is also clear that both between the wars and afterwards there were quite distinct regional variations in the amount of land being taken for urban and related development (Best and Champion, 1970). These programmes of urban expansion necessitated an increase in the amount of minerals required for all stages of the construction process.   Aggregate materials, limestone and chalk, building stone and brick making materials were all required in large quantities.

In addition to continued land take for urban expansion, the post war period has also been characterised by large scale city centre redevelopment schemes, the New Town programme and massive investment in Motorway construction and other communication and industrial projects.   Although there has been a reduction in the amount of new developments taking place in Britain during the past five years, there is a general expectancy that in the coming five year period the country will emerge from the current recession and urban expansion and new capital projects will increase.

All urban development requires specific mineral supplies. It has already been noted that the rate of urban growth in the post war period has not been evenly distributed and there is no reason to assume that the bulk of future urban growth will result in anything other than continued concentrations into particular regions. As these regions commence upon new programmes or urban growth it is clear that they must exert increased pressure upon those mineral resources essential for construction. Given the relatively high cost of transporting bulk, low value materials over large distances there are certain economic advantages in securing basic raw materials as close to the sites of development as possible. It is clear that certain regions can no longer guarantee meeting all their mineral requirements from their own resources. To attempt to satisfy specific regional demand from regionally worked minerals would, in some cases, require hitherto sacrosanct areas, such as high grade agricultural lands and Areas of Outstanding Natural Beauty, to be exploited.

Housing statistics (issued by the Department of the Environment) give an indication of new building taking place in England and Wales. In the period from 1973 to 1977 there was a decline in the number of new house starts in both the public and private sector. For the country as a whole the decline was of the order of 17 per cent though this hides a considerable regional variation from as little as a 5 per cent reduction in the West Midlands to a much greater decline in the Northern Region of 29 per cent. This trend has continued to the present time. Although data for aggregate production is not complete for all regions it is clear that in the period between 1973 and 1978 (Department of the Environment, 1978, 1980) there was a reduction in the volume of aggregate material produced.

A number of factors affect the rate at which new house building is likely to proceed, not least of which is the health of the economy, but other elements have an influence such as an increase in the formation of new households and the availability of land. Government predictions about the speed of economic recovery in the United Kingdom are more optimistic than those of the Organisation for Economic Cooperation and Development, for example, but it is anticipated that by 1985 demand in the United Kingdom economy will have made a partial recovery. By 1985 it is anticipated that the population bulge of the 1960's will have worked its way through to the stage of household formation and will thus be creating an increase in demand for new houses. In anticipation of such demand it would appear that government is trying to speed up the rate at which land can be made available for new housing.(Department of the Environment, 1980 (b); Tyne and Wear County Council, 1981).

If there is to be a period of urban expansion in the second half of the decade then there will be pressure on both land for development and for mineral resources. As in the past it seems as though there will be conflict between mineral operators, land developers, other land users, the public and local authority planners. Although it is not possible here to examine all the conflicts which are likely to arise in the coming years as a result of new and expanded minerals operations, an attempt is made to identify some of the key areas of conflict which have arisen in the past as a result of the pressure for new land for development including mineral extraction.

It is clear from the Regional Aggregate Working Parties reports (Department of the Environment, 1978, 1980) that not all regions, or even areas within regions, are capable of satisfying their own demand for minerals despite a period of reduced urban growth. If the level of urban expansion does increase then in addition to the problems already mentioned conflicts brought.about by the inter and intra regional movement of minerals will have to be resolved. There is mounting evidence to show that the increased traffic flow, particularly on roads, generated by mineral working is causing concern to planners and communities alike. (Peak Park Planning Board, 1976). Given the greatly reduced rail network in the United Kingdom it seems inevitable that more and more aggregate minerals will be transported by road. This transport problem also raises fundamental questions concerning the logic of exploiting large volumes of bulk, low value products and transporting them over uneconomic distances to areas which some would describe as being already overdeveloped. It also raises questions about the exploitation of minerals from areas which benefit little from the creation of new jobs or from value added, yet pay a high price in terms of environmental disruption.

AGRICULTURAL DISRUPTION

Mineral exploitation should not be evaluated purely in terms of its own economic significance. There are other groups in society who find their own legitimate, and often long standing, economic activities threatened by proposed mining operations. Where these would be incompatible with existing land uses, there is little hope of finding a planning solution which would satisfy those who have fundamental reasons for opposing any new development. The difficulty of finding an appropriate planning solution is further compounded where the land uses in question are, either explicitly or implicitly, supported by different government policies. Confrontation between mining and agricultural interests highlights these problems.

The impact of both deep and opencast mining on agricultural land can be severe. Although it has always been appreciated that opencast mining would have an immediate and obvious effect on agricultural production it is becoming increasingly apparent that deep mining also can be particularly harmful. Such views were expressed forcefully by the agricultural lobby at the Vale of Belvoir Inquiry. It should also be appreciated that the impacts of deep mining can be more insidious than those of opencast working and exert a far more detrimental effect on agricultural production over a longer period of time and be costly to rectify.

In the process of opencast mining agricultural production is halted. Although phased reclamation work can return land to production long before the entire opencast operation has finished, it would be fatuous to suppose that large scale opencast operations, whether they be for coal or sand and gravel, do not have a most noticeable impact upon agriculture over a long period of time. Such impacts can be measured in terms of loss of production, disruption to agricultural communities and related activities and loss of amenity. As far as opencast coal mining is concerned, despite the provisions laid down in the Opencast Coal Act of 1958, there is still a belief in certain quarters that drainage and fertility factors remain a problem following some opencast reclamation work. (Council for Environmental Conservation, 1979). Although such reports are critical of some of the practices of reclamation it is important to stress that the Opencast Executive of the National Coal Board have made substantial improvements in reclamation technology during the past twenty years and though there might on occasion be some justification for criticising their work, it is evident that without the careful supervision of the Opencast Executive during all phases of mining operations, the impact upon agriculture and amenity could have been far worse and longer lasting.

Where sand and gravel or coal, capable of being mined by opencast methods, underlie good quality or well managed agricultural land, there must always be concern that inherent fertility will be lost despite carefully controlled reclamation work. In some instances there is concern that no reclamation will take place at all, particularly with sand and gravel workings, thus reducing the national acreage of better agricultural land. Although there is evidence that many sand and gravel workings were once simply abandoned(Oxenham, 1966), during the post war period there has been a gradual improvement in the number of such sites reclaimed. Increasingly, planning authorities have used their powers to attach reclamation and aftercare conditions to permissions to extract minerals.

Urban uses, including mineral extraction, have destroyed substantial areas of higher grade agricultural land in the post war period (Best and Swinnerton, 1974). The idea of using better grade agricultural land for mineral extraction is contrary to government policy (Department of the Environment, 1976) though it is accepted that under certain circumstances this may be permitted. It would seem reasonable to assume that if the production capability outlined in two critical White Papers (Ministry of Agriculture, Fisheries and Food, 1975;1979) is to be realised then it should be of paramount importance to prevent good quality agricultural land being taken out of production. Even though in some cases it is possible to restore such land to its former quality it can be a very costly procedure and it may take many years to achieve. Thus, conflict between the need to make mineral resources available for development and industry on the one hand, and the need to retain good agricultural land in production on the other, is an area to which greater attention must be given.

Although farmers disrupted by opencast operations are likely to be compensated (National Coal Board, 1957) it is important to draw a distinction between owner occupiers, tenants and managers. The latter two groups are not as fully compensated for their disturbance if planning permission is granted for mineral workings and the land is sold or compulsorily purchased. In any event, many farmers would most certainly prefer to retain their farms intact and resist any attempts to have the land used for mineral working. This became evident at the various public inquiries in 1970 and 1975 which dealt with the National Coal Board's application to develop an extensive area of land at Butterwell in Northumberland. There are important strategic arguments to be advanced in defence of certain grades of agricultural land being taken for mineral extraction, but invariably strong emotive reasons such as 'way of life', 'tradition and community spirit', which are less easy to quantify, are also put forward at public inquiries.

The hostility which agricultural organisations have shown towards deep mining proposals is typified by recent events at the Vale of Belvoir Inquiry. Although the National Coal Board must have anticipated a vociferous outcry from amenity groups, they may well have been surprised at the strength of feeling expressed by the farming community and their representatives. This is, on closer examination, not surprising given the range of problems that deep mining can pose for the agricultural industry.

Subsidence is one of the most common features associated with deep mining and although the effects of subsidence often tend to be associated with urban areas there can be little doubt of the impact that it can have upon agriculture (Coleman, 1955). This phenomenon has been thoroughly investigated (Blunden, 1975) and advances are being made through research to find better means of predicting and preventing the occurrence of subsidence (Down and Stocks, 1977). Subsidence also affects agriculture through damage to farm buildings and land. Although it is possible for farmers to claim compensation as a result of damage incurred through subsidence, under the provisions of the Coal Mining (Subsidence) Act, 1957, the procedure for establishing liability is not always straightforward, though it is important to stress that the National Coal Board have an excellent reputation for the promptness with which they deal with claims for compensation.

There is a belief in the farming community that the compensation paid by the National Coal Board for damage to land and property does not always fully cover the cost of the losses which are incurred (Wiles, 1979). It is never easy to calculate farmers' losses where the effects of subsidence have caused delays in production or have resulted in a reduction in yield due to a deterioration in the quality of the land. Severe cases of subsidence can drastically affect land quality through alterations to the level of the water table. Unexpected costs may also be incurred. For example, if a water main breaks due to subsidence the National Coal Board will pay for the repair of the pipe, but the farmer may be obliged to pay for the water wasted (Smith and Huggett, 1979).

Deep mining can take a large amount of land out of agricultural production and although this might not necessarily be viewed as significant in national terms it can be crucial for particular enterprises. Land is required for pit head building, communications, waste tipping and coal storage. Various estimates have been made as to the amount of land which would be taken out of agriculture if the Belvoir coal field were to be developed. At the Inquiry it was suggested that 823 hectares (of which 810 hectares were in agricultural use) would be required for mineheads and tipping (Leivers, 1980). It was further suggested that additional land might be required for tipping amounting to between 166 and 336 hectares during the final third of the 75 year proposed life expectancy of the coal field. A further 100 hectares for tree planting and 250 hectares for other requirements such as new housing might well be required. Although some of the land would, in the fullness of time, be reclaimed and restored to agricultural use the net loss of land would be considerable.

The potential problems and areas of conflict associated with deep mining appear endless. Included in a list of worries that the farming community have about developments like those proposed for Belvoir are noise, dust, pollution of water courses, increased traffic flow and an increase in the range of problems associated with farming close to urban areas (Dent, 1979). Whatever might be the outcome of the Vale of Belvoir Public Inquiry it is clear that much time will be devoted, by agricultural organisations, to monitoring the problems generated as a result of mining at the new Selby coalfield. Should the National Coal Board press ahead with proposals to develop other coalfields in Cheshire and Warwickshire they are likely to meet determined opposition from those with a vested interested in agriculture.

Although legislation such as the Agricultural Land (Removal of Surface Soil) Act, 1953, the Coal Mining (Subsidence) Act, 1957, the Opencast Coal Act, 1958, the Coal Industry Act, 1975 and the National Coal Board's Code of Practice, 1976 have all gone some way to producing a framework in which the worst excesses of disruption to the agricultural industry can be reduced, it is in the nature of opencast and deep mining that large scale disruption will occur. In West Germany the impacts of large scale opencast mining have, to an extent, been aleviated by comprehensive legislation (Olschowy, 1971) and through generous cash payments or replacement farms (Dodd, 1979). Although members of the farming community in Britain might be attracted by equivalent schemes there must surely be a case for central government to coordinate the operations of the Department of Energy, the Ministry of Agriculture and the Department of the Environment to produce a long term strategy for energy, agricultural and the environment.

AMENITY VERSUS ECONOMIC GROWTH

In the previous Chapter it was made clear that the development of mineral resources was likely to continue unabated throughout the world with new reserves being sought, existing workings extended and investment into the search for alternatives. In the foreseeable future it is inconceivable that many governments will place undue obstacles in the way of those national or multinational companies committed to massive programmes of mineral resource exploitation. It is, however, likely that in the more developed countries, permission to exploit mineral resources will be granted only in the face of substantial opposition from environmentalists and amenity pressure groups.

Although there are groups in society who feel that the time is right for a major change in direction, which would drastically slow the rate of resource exploitation and economic growth, many environmentalists probably accept that the economic system is not going to change in any radical way and that their energies should be directed towards persuading governments to become more rational and careful in their approach to the use of the world's natural resources. In the fervour of the 1960's and 1970's many environmentally conscious groups and individuals were attempting to identify the hazards of uncontrolled and undirected economic growth, (for example, Commoner, 1971), as well as the short, medium and long term diseconomies of ignoring environmental constraints upon development (for example, Mishan, 1967; Bain, 1973).

In the United Kingdom a strong environmental lobby evolved in the inter war period, (there was even an Amenity Group of Members of the House of Commons) (Sheail, 1981) and it has become increasingly organised and articulate in the post war period (O'Riordan, 1976). Groups such as Friends of the Earth have campaigned nationally and internationally across a broad front on matters of environmental and economic controversy, whereas others such as the Friends of the Lake District have been content to exert pressure at a local level for the protection of amenity and environment. Given the presence of these campaigning bodies and the legal requirement to inform the public and involve them in the process of project appraisal, it seems that few major applications to develop mineral resources will pass uncontested.

Although the post war period in Britain has been a time of rebuilding and economic growth, notwithstanding one or two periods of recession, government has appreciated the need to give due consideration to environmental and amenity issues. The planning acts, national park and countryside legislation (1949,1967, 1968), the new Wildlife Countryside Bill (1980) represent some of the ways in which government has tried to improve and conserve the amenity and ecology of the countryside, whereas specific environmental legislation such as the 1973 Water Act and the 1974 Control of Pollution Act have attempted to control the problems generated by industry and this includes the range of extractive industries.   In all its attempts to introduce legislation which aims to protect the amenity and ecology of the land, government is treading a very difficult path because, however vociferous might be the voice of the conservationist, the voice of industry and commerce is just as strong, orchestrated through organisations such as the Confederation of British Industry and in appropriate cases, the Trade Union Congress.   The attitude of government in the post war period to the conflict between amenity and economic growth would appear to have been one of compromise, both in its legislation and the interpretation and enforcement of such legislation through the planning process.

Some of the dilemma which planning faces in trying to resolve the conflict between amenity and the needs of the economy is highlighted by the issue of mineral extraction in National Parks.   In the Peak District National Park carboniferous limestone exists in abundance.   This has been worked for the chemical industry, the agricultural industry and roadstone for many years and certainly long before the area was designated a National Park in 1951.   When the Peak Park was established the boundary was drawn in such a way as to deliberately exclude large areas of mineral working, particularly in the area around Buxton.   It was made eminently clear by Silkin (1949) that, whilst supporting the need for National Parks and the need to protect them from inappropriate development, where mineral wealth was in the national interest it would have to be worked.   It was in the light of this statement (plus the desire to exclude unattractive areas of land from the National Park) that the boundaries were so carefully drawn.   Although many of the larger limestone quarries and potential areas of new working were excluded from the Peak Park, it was not possible to avoid including all such areas.

Throughout the 1970's pressures were being exerted upon the Peak National Park Authority to relinquish more land for limestone working and it is clear that in areas like Hope and Eldon Hill, both within the National Park, limestone quarrying has expanded.   In November 1973, Imperial Chemical Industries submitted an application to quarry limestone in an area known as Old Moor which lay to the east of existing quarries in Great Rocks Dale and straddled the National Park boundary.   The area of the proposed mining operations came partly under the planning jurisdiction of Derbyshire County Council and partly under the jurisdiction of the Park Authority.   On the basis of discussions which had taken place between I.C.I., Derbyshire County Council and the Park Authority during the previous five years about possible ways of limiting the impact of any development at the Old Moor site, Derbyshire County Council were prepared to grant planning permission for that part of the site which came under their planning control.   However, the Peak Park Planning Board refused to grant planning permission.

Imperial Chemical Industries appealed against the decision. In 1978 the Secretary of State determined this appeal finding in favour of the appellant but subject to certain conditions. Unhappy with the decision of the Secretary of State, the Peak Park Planning Board took the matter to the High Court where it was decided that some of the conditions which the Board had originally proposed were not, as the Secretary of State had suggested, ultra vires. The outcome was that the development of the Old Moor site should be allowed to take place, but subject to detailed conditions relating to the mode of operation, rates of extraction and landscaping (Harrison and Machin, 1981).

A number of important issues emerge from this example of limestone working in the Peak District. It is likely that the Secretary of State for the Environment would have found in favour of the National Park Authority had it not been possible to justify new minerals operation on the basis of the conditions stipulated in the 'Silkin Test'. Whilst in no way wishing to question the evidence presented by I.C.I. and their consultants at the Public Inquiry it is a fact that a substantial body of evidence was presented by the National Park Authority, its consultants and others interested in preventing the development, (for example, Friends of the Earth, 1976) which came to a different set of conclusions. Not all these counter arguments were based on broad issues of amenity but some on the basis of technological assessment. For example Ley (1976) concluded that if the basis of the appeal was, that the development was absolutely necessary in the public interest and that there were no practicable alternatives, then the original refusal of planning permission should be upheld. He demonstrated that whilst some development was necessary to ensure a supply of essential minerals to lime and chemical industries, the scale of development proposed with an extension in to the National Park could be deferred until the end of the century. Equally he concluded that there were alternative methods of ensuring the supply of limestone to the chemical industry. The question which this conflict in the evidence presented to the Inquiry prompts is, should not the government be carrying out its own independent investigations into issues of supposed national need. The question of the government becoming more actively involved at an earlier stage in the planning process is developed in a later part of this book.

Whilst there were clearly issues of national significance associated with the application to develop new limestone quarries in the Peak District, there were other factors involved as well. Derbyshire County Council must have been acutely aware of the high levels of unemployment in the High Peak and their initial decision to grant planning permission to I.C.I. must certainly have been influenced by the prospect of retaining jobs if not actually generating new ones in quarrying and associated activities. The question of jobs in areas of high unemployment has obvious political and economic overtones and planning has to tread a very difficult path in attempting to establish a sensible and politically realistic balance between conserving amenity on the one hand and ensuring a reasonable level of employment opportunities on the other. This is not simply a problem for areas like the Peak where quarrying was an important activity before the creation of the National Park, but also in areas such as the Vale of Belvoir where miners in the declining Leicestershire coalfield feel that amenity considerations should not be allowed to stand in the way of new jobs being created.

The extent of the conflict that exists between those representing the particular views of amenity and economic growth depends, in many cases, on the scale of the mineral operation that is proposed. Clearly where areas of high landscape value are concerned, the confrontation is likely to be more hostile with the protagonists drawing upon the assistance of consultants and national organisations and drawing wherever possible upon stated government policies. It should not be forgotten that planners are also involved in processing all applications for new mineral workings (with the exception of coal mined by opencast methods), no matter what the scale of operation and no matter what the location. The need to give due regard to issues of amenity and economic growth are equally important even though these are of local rather than regional or national importance.

Many of the complaints about existing or proposed mineral winning operations relate to a loss of amenity which can in part be rectified. For example, although noise, dust and vibration will persist until mining operations cease, attention to detail such as earth mounding, screening and the specific routing of traffic, can reduce the impact of mining on individuals and communities. In a similar manner dereliction can be avoided by phased reclamation and landscaping. Even where large scale mining is intrusive, it is possible to lessen its visual impact and ultimately restore the landscape to its previous character (Tandy, 1973). It would be fatuous to pretend that where mining was to continue for many years local inhabitants would not suffer loss of amenity.

There are situations where mineral extraction will cause an irreparable loss of amenity. This is amply demonstrated where features of historic and archaeological interest are destroyed. In the south east of England there are many sites of archaeological interest located on former river terraces. Extraction of aggregate material from such sites means that many features will be lost. In times of economic recession it becomes increasingly difficult for planners or appropriate government departments to pursue a policy of preserving these historic sites in the face of mounting pressure for minerals for industry and development (Miles, 1980). There is a possibility of utilising legislation such as the Ancient Monuments and Archaeological Areas Act, 1979, to protect some of the nation's heritage areas from mineral winning operations (Griffiths, 1980).

If the working of certain minerals is deemed to be in the national interest then it would seem logical to provide a better framework for decision making. Situations of direct conflict generate protracted hearings or public inquiries which further delays the point at which the flow of important minerals to industry can be guaranteed. There is surely a need to amend the process by which differences of opinion can be resolved. Compromise will perhaps continue to be one means by which this end can be achieved but this does not appear to be a particularly efficient way of planning for minerals as it implies operating on a site by site basis and ignoring the opportunities offered by a strategic planning approach.

DERELICT LAND RECLAMATION

It is not the intention here to examine post war derelict land reclamation in great detail, either from the point of view of technical achievement or the legislative background. The theme is introduced because it represents one means by which the government, planners and developers have helped to overcome some of the worst fears of other land users, amenity and environmental groups, in the face of proposals to develop both small and large reserves of minerals.

There is sufficient evidence from the inter war and immediate post war periods to justify the concern felt by those opposed to mineral extraction. Extensive areas of land were worked and left blighted with no one prepared to accept responsibility for their reclamation. However true it might be that a more environmentally conscious people became increasingly critical of a number of issues, including the problem of derelict land, in the post war period it seemed to take the tragedy of Aberfan to stir government and many local authorities to action. The promptings of organisations such as the Civic Trust about the state of dereliction in the United Kingdom and its potential if reclaimed (Civic Trust, 1964, 1970), research into the technicalities of land reclamation (for example, Hilton et.al., 1967) and government legislation and funding all made a significant contribution in the nation's attempt to come to terms with the derelict land problem.

Much of the derelict land problem was either inherited from the period before the introduction of planning legislation or was created by the activities of organisations such as the National Coal Board much of whose operations were not subject to normal planning controls. Estimates of the amount of derelict land that existed in the United Kingdom in post war Britain have varied considerably largely due to the somewhat restricted definition of derelict land introduced in 1964. The six categories of seemingly derelict land which were excluded from the government's official statistics would, had they been included, have added greatly to the estimates of land lying derelict (Wallwork, 1974). Clearly not all this land had been made derelict as a result of mineral winning, but most certainly it had made a contribution. Whilst it was deemed important to find a solution to this backlog of derelict land, it was seen to be just as important to prevent history repeating itself by permitting new mineral workings to be started which were not adequately covered by conditions relating to after treatment.

Finance was the crucial issue which surrounded the backlog of inherited derelict land. There was no shortage of technical information about reclamation and some authorities, such as Lancashire County Council, had operated quite effective reclamation programmes throughout the post war period. For many authorities though, the financial burden of embarking upon land reclamation was prohibitive. The National Parks Act of 1949 had made some provision for government assistance to be given for reclamation, but this only applied to a relatively small area of the country and excluded the vast majority of badly affected areas. There were two important acts of parliament which made more money available for reclamation work in the 1960's and 1970's, notably the Local Government Act, 1966 and the Local Employment Act, 1972. Varying amounts of grant aid were made payable under this legislation and eventually some 100 per cent grants became available in December 1975 (Department of the Environment, 1976).

As far as new mineral operations are concerned it is generally accepted that derelict land problems should and can be avoided by attaching appropriate landscaping and aftercare conditions when granting planning permission. Not only is this an effective means of preventing dereliction but it is also seen as a means of ensuring that the developer pays for reclamation out of the profit of his development. This idea was developed in the United Kingdom as early as 1951. The provisions of the Minerals Workings Act ensured that those engaged upon ironstone working paid a sum of money into a 'community chest' for each ton of rock they extracted. Landowners and others deriving royalties from these mining operations were also obliged to make a contribution to this fund and the Exchequer also made a small donation (Oxenham, 1966). The mineral operators were obliged to carry out restoration work and they were allowed to draw upon the fund to finance the work but only where the cost of reclamation exceeded a prescribed cost per acre. In other words the operator had to bear the brunt of the reclamation costs himself.

Existing and proposed planning and mineral legislation places some emphasis on the use of conditions in planning permissions. Appropriate government advice to planning departments (for example, Ministry of Housing and Local Government, 1968) suggests that;

'The following tests are suggested for deciding whether to impose a condition. Is it:

a)   necessary?
b)   relevant to planning?

c)   relevant to the development to be permitted?
d)   enforceable?
e)   precise?
f)   reasonable?'

There is every indication from this type of guideline that the appropriate government departments were,and still are keen, that planning authorities should be sensible in their use of conditions. Excessive dilligence in the use of conditions could prove counter productive if the applicant felt them to be unreasonable and as a consequence appealed to the Secretary of State. Examples of the types of conditions that have been attached to minerals planning permissions are documented, (for example, Harrison and Machin, 1981).

Although local authorities do exercise their rights and use conditions when granting planning permission to mineral operators, many of them attempt to develop a dialogue with the applicants so that conflict can be minimised by discussing the nature of the problems likely to arise from a particular mode of operation. It would be wrong to assume that such an approach will overcome all problems, but if there can be genuine cooperation between applicant and planning authority it is likely that sensible and speedy progress can be made. For example, discussions between Hoveringham Gravels Ltd., Nottinghamshire County Council and the Trent River Authority amongst others, produced the basic groundwork for the National Water Sports Centre at Holme Pierrepont (Briggs, 1971).

Conflict and partnership present methods of reducing conflict. The Project Appraisal for Development Control team at Aberdeen University stressed the need for joint working between local planning authorities and developers, (Clark, Chapman, Bisset and Wathern, 1976). Even with the best possible working relationship, conflict cannot be fully avoided. It can, however, be minimised through robust and explicit strategic planning which is based upon political resolution and unequivocal policy.

SECTION III

DECISION MAKING, EVALUATION AND PLANNING.

6 Influencing Elements

As has been demonstrated in the preceeding section, a wide range of factors influence the process of minerals planning. All too frequently planners claim to have little or no influence over the form and operation of private and public sector economic activity; in the case of mineral resources there is clearly a direct relationship. The close interaction in minerals planning between evaluation, policy development and decision making is therefore unusual, and possibly unique. Many organisations and agencies have a vested interest in the outcome of planning decisions on proposals for exploration and exploitation. This vested interest is a reflection of the important role played by minerals in economic activity and, additionally, reflects a general concern for the quality of the environment. Central to the arguments advanced by the protagonists at many public inquiries is the definition of what constitutes the national interest. The Stevens Committee (1976) argued that in any discussion of minerals planning 'it is almost inevitable that strong conflicts of interests will arise', and that such conflicts were 'at the roots of the problems' that they were seeking to examine. The resolution of conflict is a difficult task, for it involves planners in attempting to reconcile, firstly, the differing objectives and approaches of the various parties concerned with resource development, and secondly, the distribution of the costs and benefits which accrue from mining operations. Questions of national and in some cases international policy are also involved.

No discussion of the factors which influence the strategic planning of mineral resources can afford to ignore this essential question of national interest. Resource planning is essentially concerned with future needs and the availability and continuity of supply. However, needs will change, existing sources of supply will diminish and new reserves may be discovered. The role of central government departments and the arrangements for national planning should take account of these changes, and attempt to accommodate them in existing or new policies at national, regional and local levels.

RESPONSIBILITIES AND INTERESTS

Although there is no single national policy for minerals planning, there are a number of central government policy areas where minerals play a prominent part. These policy areas reflect departmental responsibilities and, in certain cases, cause conflicts of interests to arise between departments.

The Department of the Environment is the sponsor for a range of minerals mainly utilised by the construction industry. In their evidence to the Stevens Committee (1976) the Department outlined their role. 'It is, first, to ensure that the needs and practical problems of the industry are taken into account in the formulation of the policy; second, to interpret that policy to the industry and to provide the industry with a clear point of entry into contact with the Government machine'. Due to the high level of national self sufficiency in construction industry minerals the Department of the Environment portrayed its role as 'evolving some sort of a strategy' and providing 'realistic advice to planning authorities as to what sort of account they should take of the national interest'.

Energy minerals – coal, oil and natural gas – are the responsibility of the Department of Energy. Clearly there is a high degree of interaction between the use of these minerals and fluctuations in world commodity prices. This relationship is expressed in the evolution of specific policy proposals for control over the rate of extraction, the encouragement of new developments and the emphasis which has been given, by the Department, to issues concerned with the national interest.

The majority of metalliferous, and many other industrial minerals are regulated by the Department of Industry. In their evidence to the Stevens Committee (1976) the Department noted their desire to encourage mineral operators through the provisions of the 1972 Mineral Exploration and Investment Grants Act. Although the Department agreed that their policies should be reflected in structure plans, little evidence was presented to demonstrate that any steps had been taken to 'ensure that the relevant departmental policies, and associated factual material, are offered to county planning authorities'. This failure prompted the Stevens Committee to suggest that the Department should adopt a more positive approach to its role in relation to mineral planning matters.

Rural and agricultural interests are represented by the Department of Agriculture who see their role as the defence of 'good quality agricultural land'. (Stevens, 1976). Through their contribution to the planning process the Department attempts to exercise influence in pursuit of this objective. As many aggregate minerals are worked at short distances from major urban agglomerations, (often in areas of good quality agricultural land) there have been cases where conflicts of interest have arisen. With the movement of coal and other mining operations into lowland Britain – into possibly the Vale of Belvoir and South Warwickshire – then these conflicts can be expected to increase in number.

A number of government agencies also have a specific interest in mineral resources. The Institute of Geological Sciences provides detailed information to planning authorities, and whilst not directly involved in conflicts, the information which is provided by the Institute is, in some cases, contested. The Health and Safety Executive has a concern with safety and hazard, during the operation of mining and following extraction. Their advice and opinion has, as in the case of proposals for the construction of hydrocarbon reception terminals, sometimes conflicted with direct commercial interests.

The interests of the operators are specified by the very nature of their activities. Restrictions on exploration and extraction, the limitation of annual yield and the imposition of severe operating and restoration conditions can all adversely affect the commercial viability of a mining operation. Whilst it is a truism, one major mining operator expresses the problem simply, 'mining operations by their very nature involve disturbance of the natural environment' (Rio Tinto Zinc, 1978). For that company the need to continue to extract mineral resources is clear, this need is expressed in terms of company performance, with profits for 1976 standing at £81·3 million, 'an all time record' whilst in the same year the environmental responsibilities acknowledged by the company resulted in an expenditure of £20 million (Rio Tinto Zinc,1977, 1978). Although this is an extreme case it demonstrates the financial interests which inevitably underpin the arguments for extraction which are advanced by operators.

Another approach which has been used by operators is to argue that their proposals are of national strategic significance. The case argued by the National Coal Board at the Vale of Belvoir inquiry was summarised by Hills, (1979). 'The energy crisis makes this self interest (of the National Coal Board in mining in North East Leicestershire) appear synonymous with the national interest, for how else will Britain survive the inevitable exhaustion of North Sea oil and gas than by utilising its coal resources?' In considering the trade-offs that frequently exist in proposals for mineral development the planning system faces a serious challenge. This problem has often been expressed as a choice that has to be made between jobs and the environment. (R oyal Town Planning Institute, 1979).

DIVERGENCE OF INTEREST AND PLANNING POLICY

Blowers (1980) in his study of the mining of fuller's earth in Bedfordshire illustrates the nature of the arguments and counterarguments which are frequently presented. Fuller's earth, from which bentonite can be produced, was seen by the applicant company and by the Department of Industry to be a vital material. The company argued that there were few areas where fuller's earth could be commercially exploited and that any failure to maintain the level of output would adversely affect the balance of payments. This view was supported by the Department of Industry, who in a letter to the planning authority stated that they 'hope that the council will be aware of the national interest when deciding upon this company's application'. Local objectors claimed that the demand for the product was falling and that due to the use of part of the existing production for the manufacture of cosmetics and cat litter, 'if the Government is serious as to this substance being of strategic need and in limited supply, it would hardly allow 35,000 tons to be used in this way'.

A point of interest is that, in Blower's view, it appeared that all the experts on fuller's earth were already employed in the industry, and that the planning authority were unable to retain an expert witness who could comment on the issue of strategic need.   In the event, the county planning authority refused permission for the extraction of the mineral.   At the subsequent inquiry the various cases were presented and resulted in the decision of the Inspector to recommend'that the appeal in its submitted form be refused'.   In his summary statement the Inspector noted that, 'I appreciate the importance attached to this comparatively scarce mineral and its role in the national economy, but I am of the opinion that the effect of the excavations upon the local community and the landscape would be greatly ameliorated if the operations were to be made in the reverse direction'.   Following the Inspector's report further observations were allowed from the interested parties and finally the Secretary of State allowed the appeal subject to twenty detailed conditions.

The case of the fuller's earth in Bedfordshire illustrates the extreme complexity of the arguments which are advanced when matters of national interest are at stake.   It also demonstrates the divergence that frequently occurs between the national interest and local interest.   Expressed in simple terms, this divergence is often typified as national economic interest against local environmental interest.   This simple relationship does not always hold true.   In the case of coal mining, mineral extraction is a significant regional and local source of employment.   As can be seen from table 6·1, the extraction of coal provides a

Table 6·1

Regional Employment in Coal Mining 1976

Region	Number Employed*	Proportion of Total Employment**
South East	5·4	0·07
East Anglia	–	–
South West	–	–
West Midlands	23·4	1·07
East Midlands	66·1	4·42
Yorkshire and Humberside	78·9	4·01
North West	12·3	0·47
Northern	45.7	3·64
Wales	37·9	3·81
Scotland	27·8	1·34
Great Britain	297·5	1·35

* in thousands   ** as a percentage of total employment
Source :   Department of Employment, (1977)

70

major source of employment in the East Midlands, Yorkshire and Humberside, the North and Wales; in addition a large number of coal related jobs are provided through the multiplier.

The economic and environmental conflicts noted previously do not always hold true, but in certain cases a satisfactory and almost symbiotic relationship can exist. In the case of the Amey Roadstone Corporation's gravel workings at Sutton Courtney, (Oxfordshire), extraction is followed by progressive restoration, using refuse from Greater London. In this instance the waste disposal problem facing a major urban area provides the fill material required to achieve the satisfactory restoration of thirty acres of land per annum. Although this case is by no means typical of all mineral workings, it does illustrate what can be achieved given mutual rather than divergent interests. (Amey Roadstone Corporation,1980).

From a planning standpoint the interests of national government departments, operators and local residents are all significant factors that have to be accommodated within strategic policy. At county level the structure plan provides an opportunity to integrate minerals issues with other economic and environmental considerations. Furthermore, the obligation placed upon county authorities to prepare and submit waste disposal plans allows the further extraction of minerals to be linked to proposals for restoration. The preparation of subject and local plans, in accordance with approved structure plan policy, can enable the planning authority to attempt to resolve conflicts between national and local interests. Whatever planning solution is generated, it almost inevitably results in a compromise; the dilemma is a common one, that is, that a simple distinction between national economic interest and local environmental concern is not sophisticated enough to allow for the diversity of local interests that exist in any area.

In some regions the national and local interests can be seen to be compatible, as such, structure plan policy is intended to ensure the effective extraction of minerals which are fundamental to local activities and which meet national requirements. This is clearly the case where basic industrial activities are dependent upon a limited number of sources of a mineral, where undue restrictions upon the extraction of that mineral would adversely affect, either the total level of production, or the costs of the associated industry. In the Merseyside Structure Plan (1979) the significance of Shirdley Hill sand, an important raw material for glassmaking, was recognised and it was recommended that, 'the extraction of sand for special industrial uses will normally be permitted within defined consultation areas'. The consultation areas, adjacent to the glassmaking centre of St.Helens, are areas of potential reserves of the sand where local planning authorities have agreed to inform the glass industry of development proposals. No economic alternative to the Shirdley Hill sands exist, thus local and national interest are not in conflict and strategic planning policy can be developed within a stable framework in order to ensure a long term programme of supply.

Whilst structure plans are willing to recognise the importance of national policy, in certain areas where employment prospects are not greatly enhanced by the extraction of a more generally available mineral, then a more cautious approach is adopted. Tyne and Wear County Council, (1979) acknowledge that although 'new deep coal mines are unlikely', it is clear that 'existing deep mines could be further developed and open cast mining of coal is an increasingly important element in coal production plans'. Accordingly, the county, in its structure plan, has sought to ensure that the production of open cast coal is restricted in order to maintain amenity. Many other examples of this approach could be provided, indeed, the majority of structure plans attempt to walk the tightrope of jobs versus the environment.

The dilemma that faces county authorities is hightened by the absence of explicit national and regional policies towards specific minerals. Whilst the present authors would agree with the Stevens Committee (1976) that there are many dangers in attempting to produce a 'national plan' for all minerals, they also support one of the conclusions reached by the Committee that, 'Every effort should be made to formulate and refine national policies for individual minerals or groups of minerals and to reconcile these policies with each other and with national policies for other land uses'. Without such guidance then the work of Regional Aggregates Working Parties and county planning authorities is made more difficult.

For many minerals the consequence of this absence of agreed policies is that national and local interest is decided through the appeal and public inquiry procedure. Although the inquiry procedure is essentially the final stage in the process of decision making, it is of specific significance in minerals planning, because the evidence which is submitted at many appeals is based upon claims that the proposed extraction of a mineral is in the national or regional interest. The validity of this argument can be, and frequently it, questioned and (whilst not wishing to question the role of the Planning Inspectorate) can exert influence upon the Secretary of State in making a decision upon proposed developments.

NATIONAL INTEREST AND DECISION MAKING

As was shown in the example of fuller's earth in Bedfordshire many proposals for the extraction of minerals are considered at public inquiries. It is therefore necessary to examine the extent to which arguments of national interest are used at inquiries, and the degree of influence that such arguments exert over the final outcome. One of the major difficulties that is encountered in using the public inquiry to determine the validity of an application for the extraction of minerals is that the real issue can be avoided. Given that both the development control and public inquiry systems are required to determine a specific application, then broader policy questions are, in theory, subsidiary to that role. However, in practice the determination of an application and the rationale that underpins the application, become interwoven. The Stevens Committee (1976) summarised this dilemma, and noted that the real issue in many inquiries is, 'Should minerals be worked at the place covered by the application being inquired into, or somewhere else?'. It can also be added that the question should be asked whether minerals should be worked at all. A fuller discussion of the inquiry system is presented in Chapter 8 of this text; at this stage attention is focused on the definitions of interest which have been presented at inquiries and the outcome of those expressions of interest.

Uden's comprehensive and excellent study of the Drumbuie inquiry (Uden, 1976) illustrates many of the issues involved in mineral resource development. The proposal, by Taylor Woodrow and Mowlem, to develop a site at Port Cam for the construction of deep water concrete production platforms encountered significant opposition. The inquiry, held in 1973, was in many senses a landmark in British planning, because it was preceeded by an impact analysis (commissioned by the Secretary of State for Scotland) which was intended to examine the effects of the proposed development. A broader context for the inquiry was provided by the work of the Scottish Development Department, who were attempting to provide guidelines for oil related developers concerning the extent of demand for oil platforms and the location of construction yards. The Department of Energy (and indeed central government as a whole) also had a specific interest in the Drumbuie inquiry, in that they were concerned to see a rapid development of North Sea oil resources. On the surface, the applicants could therefore cite government policy (for the rapid depletion of oil reserves) as supporting their case.

The applicants presented their case as being in the national interest; and argued that the Port Cam site 'has advantages for constructing oil platforms which are not found in combination elsewhere in this country'. This case was further reinforced by the applicants who also argued that as well as generating a demand for between £6 million and £7 million worth of orders from Scottish suppliers for each platform which was constructed, the project would create some 1200 jobs within South West Ross. To an extent the view expressed by the applicants was undermined by a witness, representing the Department of Energy, who stated that Norwegian platform construction yards could, in part, satisfy the United Kingdom's demand for platforms. The applicants subsequently in their presentation gave less attention to the national balance of payments arguments. However, the Department of Energy also gave evidence that government policy was 'to extract oil as quickly as was commercially practicable and to ensure through the Offshore Supplies Office that British industry got as much of the business of supplying equipment and services for North Sea oil as possible'.

Furthermore, the Department stated that there was a 'real risk that if Drumbuie was not available oil would flow less readily'. The applicants thus based their case on the following argument; given that it was in the national interest to operate a policy of rapid depletion of the North Sea oil reserves, therefore concrete platforms were required and that Port Cam was the most suitable location for the construction of such platforms. In addition the construction of platforms would provide a valuable boost to the local economy.

The objectors to the application also argued on grounds of national interest. The National Trust for Scotland (who were through a bequest the trustees for the Port Cam site) argued that it was in the national interest that North West Scotland should not be subjected to major industrialisation. Ross and Cromarty County Council objected to the proposed development 'in principle and in particular because they considered that the extent of the proposed development was incompatible with the existing infrastructure of the South West Ross peninsula'. A broader question of national interest was also presented by the National Trust for Scotland, who argued that the land had been declared by Parliament to be inalienable, and that to take the land for development would set a dangerous precedent and discourage further donations to the Trust.

After 43 days the inquiry was concluded, the Reporter (Inspector) undertook a site visit and then presented his report to the Secretary of State for Scotland. In this Report it was accepted that, from the applicants point of view, Port Cam was the best site for the construction of the type of concrete platform that was proposed. What the Reporter did not accept was that the design of platform proposed by the applicants was 'an essential part of the oil extraction programme'. On the question of visual amenity, the effects upon the local economy and social structure and the pressure that would be imposed upon infrastructure the Reporter was deeply concerned. In conclusion the Reporter noted that 'I feel that far more will be lost to the nation by granting planning permission than might be lost by refusing it'. The Reporter thus recommended that the application should be refused. Although the Secretary of State for Scotland agreed with the Reporter 'that planning permission should not be granted', he disagreed with the Reporter on a number of counts concerning the impact of the proposal on the local economy and social structure, and upon infrastructure.

Certain lessons can be learnt from this study of the Drumbuie inquiry. Firstly, it is apparent that the argument that a development is in the national interest is open to many interpretations. Depending upon the stance taken, national interest can be cited as the basis for the view that the balance of payments is of overall significance and importance. Alternatively, it can be argued that the establishment of a precedent, that land which is inalienable should be subject to development, is dangerous. National interest has many facets, there is no clear demonstration in the policy of central government as to what constitutes the national well being. Secondly, national interest is, in certain of its guises, a transitory phenomenon; as demand for a resource, or the facilities associated with that resource, change, so does the national interest. Thirdly, in the case of minerals development, there is a difficulty in separating the general need for a resource, from the need for a resource at a specific location. Although it can be argued that the planning process can only consider an individual application, the real issue at stake is the national, regional or local need for a specific mineral. Finally, it is often the case that the costs and benefits of extracting a mineral do not coincide spatially. Whilst the national economy may gain, the local environment may suffer, or vice versa. This argument was encapsulated by one of the witnesses at the public inquiry into the proposal for exploratory drilling at Capel Hermon. 'There are far too many people who regard the County of Merioneth as some form of Indian Reservation inhabited by the Welsh equivalent of the Blackfoot, Sioux and Apache Indians. We believe that if the Government wishes to conserve certain areas to the detriment of the development of those areas then there should be compensation to the area to be conserved'. (Searle, 1975). It is clear that there is no single view of the public good.

Planning policy is therefore presented with a dilemma. In the majority of cases it cannot, to any degree of accuracy or certainty, be based upon a specified national interest. If policy, at regional or local level, satisfies the needs of one central government department, it can still conflict with the views of another. Should policy coincide with the corporate view of central government, then it can still clash with the approach adopted by developers. Even if both central government and minerals operators are accommodated, it should not be assumed that the local interest will coincide with the national interest. Mineral resource developments are clearly an issue which, as Gregory (1971) argues, may be 'necessary and desirable, but, as often as not, these developments interfere with the amenities of the locality chosen for the project'.

7 Planning Approaches

For some time before the enactment of the 1947 Town and Country Planning legislation there had been wide ranging debates about the need to control mineral winning operations through some form of planning control. Indeed, the Town and Country Planning Act of 1932 had done precisely this though there appears to be little evidence to suggest that local planning authorities were particularly keen to exercise their new powers. Probably because the 1932 Act had failed to exert any significant influence over mineral developments, in 1946 the government of the day introduced an Interim Development Order, which was added to the 1932 Town and Country Planning Act, making it mandatory for any mineral operator to obtain planning permission before commencing new work. The concern that had been expressed by government bodies and emerging pressure groups, such as the Council for the Protection of Rural England, in the inter war period was to manifest itself in the objectives of the 1947 Town and Country Planning Act. Although there have been changes in the legislation relating to the control of minerals in the post war period, the issues upon which the 1947 Act focussed attention have remained central in the evolution of new policies and new legislation. The 1947 Act also clearly placed responsibility upon the local authority for the control of mineral working.

The 1947 Town and Country Planning Act sought to achieve greater control over mineral winning operations for the following reasons;

(a) 'To ensure that mineral deposits needed, or likely to be needed, to meet future production requirements are not unnecessarily sterilised by surface development (eg., buildings or roads), and are thus kept available for exploitation as occasion demands.

(b) To ensure that the necessary rights in suitable land are made available to mineral undertakers to enable them to meet national needs.

(c) To prevent or limit the working of minerals where such working would involve unjustified interference with agricultural production or with the requirements of other surface uses, or would involve undue injury to the comfort and living conditions of the people in the area or to amenities generally, and where the national need for a mineral can be satisfactorily met from alternative sources with less detriment to the public interest.

(d) To ensure that mineral working (and the associated uses of land for plant, waste disposal etc.) are carried out with proper regard for the appearance and other amenities of the area.

(e) To ensure, wherever practicable, that land used for mineral working is not abandoned and left derelict when mineral working is finished, but is restored or otherwise treated with a view to bringing it back into some form of beneficial use'. (Ministry of Town and Country Planning, 1951).

It is clear that the legislation was seeking to achieve two broad areas of control over mineral winning activities.   Firstly, there were issues of a strategic nature and secondly, those pertaining to site development.   These were topics that had been dealt with in earlier reports such as that presented by the Committee on Land Utilisation in Rural Areas (H.M.S.O, 1942) and were to be reiterated soon afterwards by the Advisory Committee on Sand and Gravel (H.M.S.O,1948). Again it can be seen through an examination of planning permissions granted for the exploitation of minerals in the post war period that broader strategic issues as well as the more localised aspects of such development proposals have and continue to exercise the minds of those engaged on minerals planning.

It is evident from the Control of Mineral Working (Ministry of Town and Country Planning, 1951) and the later version issued by the then Ministry of Housing and Local Government in 1960 that the strategic implications of minerals planning was clearly recognised.   It is perhaps not surprising that in the aftermath of a major war strategic thinking was evident in many central government policies, and that in the economic expansion of the late 1950's government was equally conscious of the need to adopt some form of strategic doctrine to ensure that the economy was in no way starved of mineral resources vital for continued growth.   Although there can be little doubt that successive governments in the post war period have been aware of the need to give due consideration to strategic matters when dealing with mineral resources it is open to some debate as to how effectively governments have been in implementing strategic policies for the control and development of minerals.   This theme of the need for strategic planning for minerals is further developed in the ensuing two Chapters of this section and the final section of the book.

Although the principle of adopting strategic controls for mineral developments was embodied in much of the post war planning legislation, either explicitly or by implication, planning has perhaps been more concerned with the site development implications of minerals winning.   This is not surprising given the level at which most planning in the United Kingdom is carried out.   In what can be a very 'hot' political arena at a county level where new workings and extensions to existing ones are proposed, together with all the attenuated activities, give rise to considerable disquiet in local communities.   Those aspects of legislation which are more concerned with site development and associated works have provided planners with a more effective means of exercising control over the developer than with those parts of the legislation which purport to deal with strategic matters.   The way in which the legislation has addressed itself to site development issues will be outlined in this Chapter.

LEGISLATION SINCE 1945

All post war planning legislation has had a bearing on mineral development, either through particular clauses dealing with specific minerals issues or generally, through viewing minerals as an aspect of land use or development. Under the various planning acts it has been possible for local planning authorities to realise the 'Aims of Control' identified in the 'Control of Mineral Working' (Ministry of Housing and Local Government, 1960) and listed above, through the powers laid out in the 1947 Town and Country Planning Act which remained unchanged in some of the subsequent legislation. The powers of the 1947 Act were:

  (a) A survey is to be made by Local Planning Authorities of the resources and potentialities of their areas. This will cover the workable mineral resources of the area and how far they can, and ought to, meet the demands upon them.

  (b) Land may be allocated in Development Plans for the winning and working of minerals and for ancillary development.

  (c) Land, or rights in land, may be acquired compulsorily for the purpose of mineral working.

  (d) Permission may be granted or refused for the working of minerals in any particular land, for the erection of associated plant or buildings, for the use of particular land for the disposal of waste, or for other ancillary purposes.

  (e) Where permission is granted, suitable conditions may be imposed.

  (f) The erection of buildings or the carrying out of other development prejudicial to the extraction of minerals on mineral-bearing land may be controlled, restricted or prohibited by refusing or qualifying permission.

(Ministry of Town and Country Planning, 1951).

Although the 1968 and 1971 Town and Country Planning Acts brought about certain changes, particularly in respect of the nature of the surveys that had to be carried out by the local authority, in essence these powere remain substantially unchanged at the present time. The importance of the introduction of structure planning in respect of minerals was that it required local authorities to integrate minerals planning with other important strategic matters.

It is quite clear from these particular powers that local planning authorities could exercise considerable control over the winning of minerals. Even if permission was granted for development to take place the facility existed for the imposition of constraints, in the form of conditions, upon the developer. Many would argue that insufficient or unwise use was made of these powers to impose conditions during the 1950's (Tain, 1980) but if this was indeed the case it could have been attributable to a number of factors not least of which might have been the relative inexperience of planning officers in exercising their new found powers and a measure of political unwillingness to interfere overly in the development process in a hectic period of post war reconstruction and economic expansion.

In the immediate post war period the Government set up an Advisory Committee to examine the whole question of sand and gravel from the broader questions of identifying reserves to the more specific aspects of site development. This committee under the chairmanship of Sir Arnold Waters (The Waters Committee) appreciated that it was not possible to exploit minerals without causing some disruption to communities and other land uses. As such there was a need to pay particular attention to the way in which permissions were granted, and that conditions should be imposed upon such permissions which took into account the necessity of ensuring that land used for minerals development should at an early stage following completion of operations be restored to some form of productive use. The various reports of the Waters Committee which spanned a period from 1948 to 1955, gave considerable attention to the way in which the disruptive effects of development for sand and gravel could be reduced. It could well be argued that the issues raised by this committee helped to pave the way for improved methods of operation and improvements in subsequent legislation and guidelines which were concerned with aspects of environmental disruption and improvement.

What the Waters Committee failed to realise was the extent to which the scale of operations and the demand for aggregate materials would increase in the three decades since it started its work (Verney, 1978). The point is further made by the fact that apart from demand increasing for minerals, the technology of winning and restoring land has changed so dramatically during the past twenty to thirty years that planning controls attached to long standing permissions now are seemingly inappropriate and inadequate (Bigham, 1973). The Waters Committee saw the rapid drive for sands and gravel as a short lived phenomenon, an immediate post war hicup in normal patterns of supply and demand (Ministry of Town and Country Planning, 1948). Had the Waters Committee been more successful in predicting the future levels of demand for sand and gravel in the post war period it is possible that it might have had the desired effect of forcing government to adopt more effective policies for the strategic consideration of the nation's mineral resources.

In addition to the Town and Country Planning Acts there are examples of legislation being introduced to deal with the problems associated with the winning of a specific mineral. One of the earliest attempts at controlling the process of the winning of minerals was the 1891 Brine Pumping (Compensation for Subsidence) Act. In 1952 the provisions of this Act were extended and the powers of the Compensation Board were increased. As early as 1939 a committee under the chairmanship of Lord Kennet was set up to consider the Restoration of Land Affected by Iron Ore Working. Ultimately the work of this committee led to the 1951 Mineral Working Act. Undoubtedly the most significant provision of this Act was that within a prescribed area, the 'ironstone district', a levy was charged for the extraction of ironstone which was paid into the Ironstone Fund. This money could then be utilised by the developer in undertaking restoration work. This Act has since been amended but the broad philosophy of the 1951 Act is incorporated in the provisions of the 1971 Minerals Act.

Similarly, under the provisions of the 1958 Open Cast Coal Act there are specific clauses which deal with the restoration of the land and the need to have due regard to matters of amenity and natural beauty. Although open cast coal operations are not subject to normal planning controls imposed by local planning authorities, but rather by the Secretary of State for Energy, there is clear evidence to suggest that the provisions made under the 1958 Act have resulted in a great measure of improvement being made in the process of land reclamation even though some groups still feel that there is room for improvement (Council for Environmental Conservation, 1979).

There are other pieces of legislation which have been introduced in the post war period and have a definite bearing upon mineral winning operations. For example, under the Agricultural Land (Removal of Surface Soil) Act, 1953, it became an offence to move surface soil away from agricultural land if it was for sale, if it exceeded a volume of 5 cubic yards in any three month period and unless planning permission had been obtained. In view of the growing concern about the amount of agricultural land that was being taken out of production as a result of development in the late 1950's and 1960's (Ministry of Housing and Local Government, Circular 37/60) it could be argued that the provisions of Section I of this legislation had certain strategic implications, but undoubtedly one of the main points of this legislation was to ensure that following the cessation of development operations, in this case mineral working, land could be restored to agriculture.

THE LOCAL AUTHORITY ROLE

In terms of providing guidance to local planning authorities, central government has relied to a large extent on the two editions of the Control of Mineral Working (1951, 1960). The latter edition, referred to as the Green Book, has formed the main stay of such guidance for over twenty years. Although some of the guidance included within the Green Book has been overtaken by subsequent legislation it still provides a measure of guidance in the

way that local planning authorities should approach minerals planning issues.
The advice given in the Green Book is far more concerned with local and tactical
issues than strategic ones and it is likely that the Department of the Environment's
current attempts to revise and update the Green Book will not significantly alter
this given that many of the provisions contained in the Minerals Bill, 1981 are also
local and tactical in nature.

Although many of the provisions in the planning legislation prior to the 1971
Act remained, to a large extent, unaltered following the 1971 Act, for example
provisions for compensation following compulsory purchase, compulsory purchase
itself, the use of conditions when granting planning permission together with a
host of other provisions, there was an important change in the post 1971 period
which arguably could have done much to influence the way in which planning
for minerals was undertaken, particularly with reference to strategic
considerations.   The change was the way in which surveys were to be carried out.
Prior to the 1971 Town and Country Planning Act local authorities were obliged
to carry out survey work as the basis for preparing a development plan.   Although
under the old style development plans local authorities had to include data
relating to minerals, it being one of the thirteen heads for which information was
requested, and although planning authorities had statutory responsibilities in
respect of controlling digging and reinstatement works, there was no obligation
to keep the state of minerals operations under review.   Hence, from a strategic
point of view and perhaps even at a local tactical level, there was not necessarily
any requirement to collect and collate updated information relating to the extent
to which designated areas had been or were intended to be worked.   This clearly
posed problems for any body attempting to develop a strategic approach to
minerals planning.

Once the powers of the 1971 Town and Country Planning Act were in operation
the situation changed and for new style development plans the following ruling
was introduced under Section 6(1) of the Act.

> 'It shall be the duty of the local planning authority to institute
> a survey of their area, in so far as they have not already done
> so, examining matters which may be expected to affect the
> development of that area or the planning of its development and
> in any event keep all such matters under review'

Under Section 6 (3) (a) of the Act 'the matters to be examined and kept under
review' were to include;

> 'the principal physical and economic characteristics of the area
> of the authority   including the principal purposes for which the
> land is used) and, so far as they may be expected to affect that
> area, or any neighbouring areas'.

So, as Bigham (1973) concludes;

> 'the local planning authority, in conducting the required
> statutory survey preparatory to the drawing up of a development
> plan (i.e., both "structure" and "local" plans under the 1971
> Act), must ensure that information is obtained about significant
> mining and quarrying within the local planning authority area.
> They must then include in their proposed plan suitable provisions
> to cover future extractive operations'.

Although seemingly less direct in its approach to dealing with minerals issues than previous Town and Country Planning legislation, the 1971 Act did in fact require local planning authorities to do far more in respect of keeping a number of issues, including minerals, under review. Evidence that local authorities did respond to this new direction can be measured in the county structure plans that have subsequently been produced and in the local and subject plans which are now being prepared. However diligent county authorities may have been in producing detailed information on minerals for their structure and other plans the fact remains that the opportunities provided for in the 1971 Act were not fully realised either at a regional or local level. It would have been possible for local authorities to effect significant changes following on from their review powers both within their own areas and through cooperation with adjacent authorities over a wider area. During the past decade regional working parties have been established to examine problems relating to the working of particular categories of minerals, notably aggregates and hydrocarbons. As yet there is no indication that agreed and comprehensive policies for all mineral development have resulted.

The powers given to local authorities to revoke or modify planning permission were carried forward from the 1962 Town and Country Planning Act and as under the 1962 Act, the 1971 Act made provision for compensation to be paid to any person who;

> ' (a) has incurred expenditure in carrying out work which
> is rendered abortive by the revocation or modification;

or

> (b) has otherwise sustained a loss or damage which is
> directly attributable to the revocation or modification'.

Given the compensation (under Sections 45 and 164 of the 1971 Act) which local authorities might have been expected to pay had they embarked upon a major reappraisal of minerals operations within their areas, it is perhaps not surprising that many local authorities did not avail themselves of the range of powers at their disposal.

FROM WATERS TO VERNEY

Any debate concerning the need to fully review existing and proposed mineral developments must raise the question of the merit of entrusting minerals matters to local planning authorities as distinct from a regional or national body. This matter was raised and discussed at length by the Stevens Committee (1976). Clearly government has appreciated that powers to deal with minerals should be vested in county and not district authorities a position reaffirmed in Department of Environment Circular 2/81. Further attention will be given to the question of 'appropriate authorities' to deal with minerals matters in ensuing chapters but it is worth emphasising that the present authors do feel that there is a need to look at new approaches to minerals planning and share the reservations expressed by Stevens about the ability of all county authorities to deal with all matters pertaining to minerals in an effective and meaningful way.

Reference has already been made to the impact of earlier reports, particularly those under the chairmanship of Lord Justice Scott (1942) and Sir Arnold Waters (1948 to 1955) which had an influence on the way in which minerals planning legislation developed. It has also been noted that the Waters Committee Reports, whilst introducing valuable ideas concerning the way in which planners should deal with applications for minerals working, failed to anticipate the rate of economic development and associated urban expansion in the 1950's and 1960's and as such grossly underestimated the demand for sand and gravel (Verney, 1978). Neither is there any indication that local authorities or central government anticipated the future demand for other minerals which have since posed major planning problems, as for example with limestone, nor was the planning system equipped to deal with major new minerals such as offshore hydrocarbons. In the evolving post war society perceptions changed, albeit slowly, and as time went by it became increasingly apparent that there was a need for a major reappraisal of a number of minerals planning related issues.

An earlier Chapter looked closely at one aspect of this, noteably the need following the Aberfan tragedy to accelerate the rate at which derelict land was being reclaimed and the methods by which this should be achieved. There was a significant amount of governmental and private action in the field of derelict land reclamation in the late 1960's and early 1970's but from a minerals point of view this was primarily a question of either looking at inherited dereliction (pre planning control) or the methods by which more effective development control could be applied to new applications for minerals developments. What was not done at this time was to look at the need for particular types of minerals in the future and the consequences of their development. It is perhaps easy to be overcritical of the efforts of local authorities to respond to the opportunities presented by the 1971 Town and Country Planning Act but in the early 1970's two important committees were set up by the government to inquire into minerals issues. The first of these committees, under the chairmanship of Sir Roger

Stevens, had the remit of examining planning control over mineral working in general whereas the second committee under the chairmanship of Sir Ralph Verney were appointed to advise government...'upon such subjects relating to the supply of aggregates for the construction industry'.

There are a number of issues raised in the Stevens Committee Report that have a direct bearing upon the adoption of a strategic approach to minerals planning, for example Chapter 5 of their report addresses itself to 'Long Term Planning'. Stevens correctly argued for the development of national policies for individual minerals or groups of minerals which could guide the activities of local authorities. It also considers issues which are clearly concerned with controlling minerals operations within county administrative areas. In short, it is concerned with improving the means by which local county planning authorities can deal with applications for permission to develop minerals and to this end a number of the recommendations contained within the Stevens Committee Report would appear to have been accepted in the drafting of the Minerals Bill currently before parliament. It is also worth noting that in its response to the Stevens Committee (Department of the Environment Circular 58/78) the government accepted the recommendation that the 'Green Book' should be revised and thereafter kept up to date. This document is currently being revised. Additional points developed by the Stevens Committee Report are raised in ensuing Chapters and its influence on the Minerals Bill is considered in the final section of the book.

Although the Report of the Advisory Committee on Aggregates (Verney, 1976) contains material pertinent to the problems of site development and after care, for example Chapter 5 'After Treatment', its approach is more strategically orientated. In response to the report (Department of the Environment Circular 50/78) the government of the day accepted many of its recommendations but there appears to be little evidence to date which would indicate that there is a willingness or a necessity to move forward to the adoption of many of the strategic issues raised. Indeed the Minerals Bill does not adopt a strategic approach to minerals planning. The remit of the Verney Committee was to look at issues pertaining to the supply of aggregate materials but it could well be argued that many of the themes developed within its report have a wider applicability throughout the field of minerals planning.

In the post war period there have been occasions when a framework for adopting planning strategies which took account of important and related economic and resource issues almost seemed a reality. It is unfortunate that the Waters Committee did not foresee the future in a clearer light else they might have pressed for the adoption of a strategic planning framework for future mineral development. However, Verney (1976) noted that in the 1950's the various Planning Ministries did arrange a series of regional conferences to assist local planning authorities in the preparation of their development plans. Perhaps an opportunity did exist at that time for central government to talk to local authorities about the advantages of inter authority discussions in respect of securing the most rational development of mineral resources, indeed a point which is made in the Verney Report itself.

With the exception of certain clauses contained within the 1971 Town and Country Planning Act referred to above, planning legislation has been centrally concerned with matters relating to permission to work minerals at specific locations. An opportunity was created to consider minerals in the context of a regional plan, and hence influence the formulation of Structure Plan policies for minerals, during the preparation of the Regional Studies and Strategies between 1964 and 1979. Although many of the strategies make reference to minerals there is no indication, perhaps with the exception of the various strategies for the South East Region, that any of these documents were intent on producing a carefully formulated and rational strategy for the development of the mineral reserves within their regions, or in the offshore province adjacent to their region.

The one instance where there has been a conscious effort by the government to inject an air of rationality in the planning for future minerals has been in the area of aggregates. Again in the introduction to the Verney Committee Report, reference is made to the work started in 1969 of the seven joint sand and gravel working parties established in the South East Region to look at supply and demand factors within their areas, their work being brought together and assessed by the Standing Conference on London and South East Regional Planning. By 1975 Regional Aggregate Working Parties had been established in England and Wales, their role was to promote discussions about more effective ways of planning for minerals in particular regions and for keeping government informed in respect of pertinent statistical data. Although two sets of statistics have been produced (Department of the Environment, 1978, 1980) the data is in part incomplete and the working parties themselves have no statutory powers to implement any regional strategy they propose without reference back to the appropriate authority, the county council.

It is not the intention of any part of this book to undervalue the improvements that have been made in the legislation and procedures for the more efficient and effective controls of mineral developments. Indeed, in the consideration of the Minerals Bill acknowledgement is made of the way in which improvements have been made. However, the present authors feel that it is time that the legislation relating to minerals planning came of age and developed an additional arm to deal with the broader strategic issues which should condition, as was anticipated through the heirarchy of plans, the way in which sites should be selected for development prior to any consideration of the way in which those specific sites should be considered at a tactical level. The following Chapters of this text attempt to provide an analysis of the failure of current procedures to deal with strategic aspects of planning for minerals and to suggest appropriate means by which this problem can be rectified.

8 Strategic Guidance and Decision Making

The central theme of this text is the need, which has been expressed in recent years by an increasing number of authors, agencies and authorities, for a strategic view to be adopted in the future planning of mineral development. Although in practice it is often difficult, if not impossible, to distinguish strategic thinking and action from other planning and development functions it is important to attempt to define and articulate what constitutes the strategic element within planning. More importantly it is crucial that strategic thinking is related to the ways in which plans are prepared and implemented. It is also important that the consequences of past failures to provide strategic plans and guidance are examined and evaluated.

Strategic planning as an activity is not the sole prerogative of central and local government, it is also an important function within industrial and commercial enterprises. In the minerals industry there are many large national and multinational firms whose concern for an individual localised deposit or working may represent only a small fraction of their corporate activities. This increasing size of firm, related to a tendency towards the diversification of their interests and the international nature of their operations, has caused many companies to plan their activities within a strategic planning framework. Whether set within the public or private sectors, strategic planning has certain common features and manifestations.

Throughout the 1960's and 1970's many forms of strategic planning were introduced into British practice. The regional plans and strategies, the strategic structure plans, the development plans of public utilities and the corporate strategies of private sector companies all have certain common objectives. They all have a strong relationship with policymaking and tend towards a long term view, they are concerned with some determination of a future state and they frequently have regard for interactions between activities. (McLoughlin, 1978). The selection and evaluation of alternative strategies is a difficult matter. Once a choice is made, on the basis of the best estimates of the consequences of a given set of actions, then to a certain extent other possibilities and options are reduced. It is therefore important in the case of minerals that the strategy which is chosen should be robust and that it should be clearly reflected in other lower order planning activities. To achieve this level of strategic agreement inevitably implies that there should be broad agreement between central government, local planning authorities and mineral operators. Once a mineral has been developed there can be no return to a previous situation.

Agreed, unequivocal, policies expressed in strategies can also aid decision making. As has been argued earlier in this text, the absence of explicit national policies has led to a number of long and tortuous public inquiries where the need for a mineral has had to be considered, frequently in isolation from other matters of policy. Furthermore, if each application for permission to develop a mineral resource is considered on its own merits then it is likely that

such incrementalism can work against other long term objectives.   This was seen
by Verney (1976) to be the case if the problem of aggregates shortages in South
East England was not resolved.   In the absence of an available supply of
aggregates then either absolute shortages would result, or imports of materials
would ensue at a higher cost, with the problem of supplying materials being
transferred to another region.   It has been argued that the consequences of not
having a strategic planning system for minerals are economic inefficiency and
environmental damage and that the public inquiry has therefore, by default,
to make strategic decisions in a policy vacuum.

This Chapter examines the extent to which minerals have been considered in
national and regional strategic plans, the role and function of structure plans
and the ways in which the planning inquiry system has dealt with strategic issues.
An assessment is also made of recent attempts at the development of strategic
guidance for mineral working.

NATIONAL AND REGIONAL STRATEGIES

In earlier parts of this text reference has been made to the fundamental role
played by minerals in the process of national and regional development.
Many United Kingdom regions owe their past and present patterns of economic
activity to the occurrence of minerals such as coal, iron ore, salt, limestone,
tin and hydrocarbons.   Both the negative and positive aspects of mining and
quarrying have been the subject of national and regional debate.

At national level specific attempts have been made to develop strategies for
the future of key minerals.   The efforts of the Waters Committee were directed
towards ensuring that an adequate supply of sand and gravel was made available
for post war reconstruction, whilst more recently the Department of Energy (1977)
has worked in co-operation with the National Coal Board in planning for the
expansion of coal production.   These recent exercises represent the latest
stages in government involvement in the strategic planning of the coal industry
(Jackson, 1974).   Such sector specific strategies have a marked regional
dimension, with clear effects upon labour demand in those regions where the
minerals occur and, in some cases, a limiting effect upon other activities, such
as the construction industry.   Furthermore, the strategic planning activities of
other government departments and nationalised industries has an influence upon
the location and extent of minerals working.   This can be seen most clearly in
the plans of the Department of Transport, where major road construction proposals
inevitably increase the pressure placed upon supplies of aggregates.   Whilst
these national planning exercises can, and do, directly affect the production of
individual minerals they tend to be less spatially specific than regional and
structure plans.

Amongst the published or approved regional strategies there are marked differences in the attention which is paid to issues related to mineral resources. One of the earliest regional strategies, that for the South East, gave detailed consideration to the need to maintain an adequate supply of sand and gravel in order to meet the demands generated by continued urban development. In the view of the strategy team it was agreed 'that an objective of the regional strategy should be to avoid sterilisation of mineral resources', but also that, 'in the longer term it will be impracticable to satisfy all demands without damaging conflicts with other countryside objectives' (South East Joint Planning Team, 1970). The emphasis given to minerals in the Strategic Plan for the South East demonstrates one important principle that can be seen in the strategic planning of resources, namely, that minerals seem only to be of concern when problems occur. By way of contrast with the attention paid to mineral resources in the 1970 South East strategy, the Strategic Plan for the Northern Region (Northern Regional Strategy Team, 1977) gave little attention to such matters, with the exception of the coal industry. Once again attention was focused on that sector of the minerals industry which presented problems. The economic and social consequences of previous patterns of coal mining, the adverse effects of dereliction and the future prospects for the coal industry were highlighted.

Other regional strategies have only considered minerals in an oblique sense. The Yorkshire and Humberside Regional Strategy Review (Yorkshire and Humberside Economic Planning Council, 1975) paid some attention to the coal industry, the report of the East Anglia Regional Strategy Team (1974) noted the recreational potential of disused sand and gravel workings, whilst the South West Economic Planning Council (1974) made little or no mention of minerals. However, in all of these strategies, infrastructure proposals are presented which have massive implications for the aggregates industry. In addition, in some regions, for example Yorkshire and Humberside, a significant proportion of the region's employment and gross domestic product is derived from mining and quarrying. The strategies noted above were all prepared or in the course of preparation before the establishment of Regional Aggregates Working Parties. They rightly identified other matters to which they had to direct their attention, even though some consideration of minerals would have subsequently proved to be of great value, especially to county planning authorities.

The only regional strategy published in recent years is that for the West Midlands (Joint Monitoring Steering Group, 1979). Despite having the advantage of being able to call upon an established Regional Aggregates Working Party, and whilst making proposals for new infrastructure that will be vastly consumptive of aggregates, this regional strategy does not discuss the question of mineral resources. The absence of any consideration of minerals issues, apart from some reference to the problem of dereliction, is all the more surprising in a regional strategy which places a major emphasis upon stimulating the economy. For example, no mention is made of the potential future role of the coal industry in Staffordshire or South Warwickshire, or the not inconsiderable significance of limestone quarrying and cement production in other parts of the region.

What is clear from this somewhat rapid consideration of national and regional strategic planning exercises is that no systematic or co-ordinated view has been taken of how and where to develop mineral resources. Neither has there been any attempt to provide explicit guidelines for local planning authorities as to what weight they should place on strategic considerations in the consideration of mineral resources. It is somewhat ironic that the Hunt Report (1969) should have given such detailed attention to the negative consequences of mineral working and the inhibiting effects of dereliction, for, from this report, evolved a co-ordinated national effort to reclaim derelict land. In the absence of national and regional strategic guidance mineral operators must look to county structure plans for advice on the development of minerals. The only exception to this general picture of gloom can be found in Scotland, where the Scottish Development Department has attempted to provide a strategic framework for the future of hydrocarbons, aggregates and other minerals.

STRUCTURE PLANS

The evolution of the recast planning system in the 1960's is notable for the attempt which was made to provide greater strategic guidance for the future development of county areas. A clear distinction emerged between the need for a higher tier of strategic plans (the structure plan) and a more detailed set of local plans (the district, subject or action area plans). Attention is focused here on the structure plan. Local and subject plans are considered in the following Chapter.

The 1968 Town and Country Planning Act (subsequently amended and incorporated in the Town and Country Planning Act of 1971) provides for a local planning authority to prepare a structure plan for its area. The local planning authority in this case is the county planning authority except in the Lake District and Peak District National Parks where special planning boards had been established. If the terms of reference for structure plans are examined then, in the case of minerals, the task placed upon county authorities was a difficult one. Matters for which structure planners should have regard are the current policies with respect to the development of the region within which the county is located. Given the absence of any meaningful regional policies for the strategic planning of minerals it is difficult to see which policies could be incorporated within structure plans. The absence of any regional guidance has inevitably forced, even adjacent, county authorities to prepare structure plans which consider mineral issues as an internal county matter, this is despite the view expressed in the legislation which requires county authorities to consider the characteristics of neighbouring areas.

Both the 1968 and 1971 Acts specified a number of matters which must be considered in the preparation of structure plans. These include population, employment and transportation, however, no specific mention is made of minerals. It can, of course, be argued that for many county authorities minerals provide an employment base, or that in the context of a specific area they are a relevant matter which should be considered. In 1974 the Department of the Environment provided additional advice to structure plan authorities (Circular 98/74). This distinguished issues of key and of structural importance. Key issues included the location and scale of employment, the transportation system and the location and scale of housing. Some other issues which may be of particular importance were also defined; conservation, recreation and tourism, shopping provision and the location and scale of reclamation, but again this did not include mineral working which in some cases is inexorably linked to questions of dereliction and restoration. The only specific comment in Circular 98/74 which mentioned mineral working related to the period of time covered by the plan. It was acknowledged that minerals planning was possible and desirable over and above the standard 15 year time horizon for the structure plans, however, no guidance was provided as to how longer term planning was to progress.

Another interesting matter concerned with the production of structure plans was raised by the 1974 Town and Country Planning Regulations. In these regulations, which set out the form and content of structure plans, it was specified that no diagram (the key diagram) was to be prepared on a map base. The arguments used to support this approach were that it would avoid creating a misleading impression of precision and would minimise blight. Whilst it is correct that for the majority of land uses these twin aims can be achieved through the use of a key diagram, in the case of minerals, which are site specific, it might be misleading and induce blight if a map base is not used. This has led some authorities to decide not to show mineral proposals on the key diagram, for example, Cornwall County Council (1980). Stevens notes (1975) an interesting solution to this difficulty, that is, to specify 'mineral consultation areas' in a report of survey rather than to compromise the production of the final structure plan. Central government was less enthusiastic about this suggestion and felt that the use of this consultation area procedure should be left to the discretion of local planning authorities.

It might reasonably have been expected that, given the difficulties faced by some county authorities in accommodating minerals issues in structure plans and given the availability of the reports of the Stevens and Verney Committees, that the revised guidance to local planning authorities issued in 1979 would have attempted to resolve some of the difficulties that have been noted above. By 1979 it was also apparent that there were important new factors to be taken into account. The experience of onshore and offshore hydrocarbon development, the difficulties of providing for new coal mining and the increasing problem of ensuring a sufficient supply of aggregates all required clearer guidance to be provided for structure plan authorities. The Department of the Environment in Circular 4/79 attempted to provide up to date guidance to local planning authorities. Again central government urged local authorities to take account of accepted regional strategies and policies, yet no specific reference was made to the work of the, by then, well established Regional Aggregate Working Parties. No reference was made to the idea of 'mineral consultation areas', or to the difficulties of representing mineral resources on a key diagram. Once more minerals were not considered specifically to be a matter of key structural importance. In all, local planning authorities were left to make their own decisions as to how they dealt with mineral resources.

The reaction to this lack of guidance from central government has taken a number of forms. Some structure plans have included a detailed consideration of minerals and associated issues, argued on the basis that such matters are relevant within the county area and that certain site specific minerals have important economic, environmental and social implications. Other county authorities have amended their structure plan policy statements regarding important minerals, the development of which are likely to produce unusual or special problems. Hampshire County Council has, for example, modified its structure plans to incorporate a number of policies relating to hydrocarbon developments (Hampshire County Council, 1981). Some county authorities have adopted a different approach, whereby minerals are considered through a local or subject plan, these plans are, as yet, untested in practice and may produce new difficulties of integration with structure plans. An alternative method of dealing with the strategic planning of minerals is through non statutory policy statements. This approach has been used for some years. Cheshire County Council (1969) developed a policy, during the late 1960's, for the control of salt extraction within the county area. More recently Cornwall County Council has attempted to provide, in co-operation with the china clay industry, a policy framework for the St.Austell area (Cornwall County Council, 1974 and 1979). Structure planning in England is not lacking in examples of good practice or innovative solutions in matters concerned with minerals, but what is absent is any significant degree of consistency or any major national co-ordination of the efforts of individual county authorities.

It is worthwhile to examine briefly the situation in Scotland. Elsewhere in this text reference has been made to the attempts of the Scottish Development Department to provide national planning guidelines for, amongst other matters, the control, development and strategic planning of aggregates and hydrocarbons (Scottish Development Department, 1974 and 1977). These national guidelines do provide an explicit context for the activities of the twelve regional and island planning authorities. Structure plans in Scotland have, like their English counterparts, given varying attention to minerals. Lothian Region has, for example , attempted in its structure plan to define policies which protect all significant mineral deposits from sterilisation and ensure that restoration is carried out to the highest standards. In addition, it has adopted the approach, also encountered in England, of working with mineral operators, the Institute of Geological Sciences and district authorities in the designation of mineral consultation areas (Henry, 1980). Regional reports provide Scottish authorities with the ability to consider a range of strategic matters related to the development of minerals. Given that the exploration and subsequent extraction of offshore hydrocarbons has been important in the recent fortunes of the Scottish economy, it is somewhat surprising that only one out of the first round of regional reports (that for the Grampian Region) gave detailed consideration to the effects of oil and oil related developments (McDonald, 1977).

Many of the weaknesses which have been identified in the strategic planning of minerals are a reflection of general difficulties associated with the structure planning system. Put succinctly, minerals issues in structure plans represent 'the interface of environmental needs and the commercial requirements of mineral exploitation' (Tain, 1980 b). There has been a tendency in many structure plans to state policy towards minerals in terms which are difficult to define or in ways which require a judgement about imprecise or non-existant national policies. In approved documents these difficulties are often compounded by the modifications which are appended by the Secretary of State for the Environment. The consequence of such imprecision is that individual applications are subject to detailed investigation, whilst many are determined at an inquiry or on appeal.

THE PLANNING INQUIRY SYSTEM

In the absence of explicit national policies for many important minerals, especially aggregate and energy resources, and in the context of frequently imprecise structure plans, it is not surprising that a substantial number of applications for mineral working are considered at a public local inquiry. There are a number of ways in which an application for planning permission to extract minerals may come before a public inquiry; these are :

i   The local planning authority may refuse permission and the applicant appeals to the Secretary of State,

ii   The local planning authority does not determine the application within the statutory or other agreed time limits and the applicant therefore appeals,

iii   The local planning authority wishes to grant permission but in such a case this would represent a departure from an approved plan. If the Secretary of State so wishes he can determine the application, thus requiring an inquiry to be constituted,

iv   The application can be called in by the Secretary of State
       in order that he might determine it (Royal Town Planning
       Institute, 1979).

Public inquiries are also held to consider and determine statutory documents,
certain of these plans involve or relate to minerals.   In the case of structure
plans the traditional form of inquiry was considered inappropriate, instead there
is an examination in public of a number of selected matters.   Local and subject
plans, after a period allowed for public participation and consultation, can be
subject to a public local inquiry.   The Secretary of State for the Environment
can require the local authority to submit the draft local plan to him for his
approval if the plan raises issues of national or regional importance, or if the
plan gives rise to controversy which extends beyond the area of the plan making
authority. (Department of Environment and Welsh Office, 1977).   In the case
of minerals subject plans it is likely that important or controversial matters will
be raised.

The more usual situation in which a public inquiry is held is where an
application for planning permission is refused or is called in by the Secretary of
State.   In such cases there are a number of different formats that can be used
for the subsequent inquiry.   Traditionally, inquiries are established by the
Secretary of State who appoints an Inspector; the appellant (or applicant),
the local planning authority and other parties make representations at the
inquiry.   The Inspector, if he is not authorised to give a decision himself, reports
to the Secretary of State setting out the facts, his conclusions and his
recommendations.   From this the Secretary of State proceeds to a decision, which
may reject the recommendation made by the Inspector (Ardill, 1974).   Two other
forms of inquiry may be used.   A two stage inquiry can be held to agree the main
issues and contentious points at a first meeting, following this an inquiry is
convened to investigate those particular points (Royal Town Planning Institute,
1979).   The Secretary of State can alternatively refer the matter to a Planning
Inquiry Commission.   A commission of this type can carry out research into the
matter referred to it as well as hearing evidence from the interested parties;
the notable feature of this form of inquiry is that it can consider issues which
relate to a number of sites located in different local authority areas.

The main criticisms of the traditional public local inquiry relate to the scope
of the inquiry and the submission of written statements beforehand.   In the
evidence submitted to the Stevens Committee (1976) both the minerals industry
and conservation interests advocated greater use of pre-inquiry statements.
Industry claimed that such statements would reduce the time wasted on
irrelevancy and repetition and would, therefore, enable the proceedings to
concentrate on key issues.   The evidence presented by conservation interests
wanted industry to disclose complicated technical matters in advance in order
that assessments might be undertaken and a case prepared.   It is certainly true

that participation frequently introduces new and often complicated evidence, or even changes the basis for their arguments whilst the inquiry is in progress. Earlier in this text the Drumbuie inquiry was considered. During the course of that inquiry a number of important changes of direction occurred in the arguments presented. It would be naive to assume that the various cases developed at a public inquiry would not be subject to change even following the presentation of previously prepared statements. However, at least such changes would be set within the context of the original cases presented. The reluctance to fully disclose information can, in some cases, be explained on grounds of commercial confidentiality or on the basis that insufficient national, regional or local strategic guidance exists upon which to fully argue for or against a proposal. This issue is linked to the problem of defining the scope of the inquiry.

Local authorities, in their evidence to the Stevens Committee, criticised the scope of inquiries. They argued that many inquiries could not examine the real issue which was,'should minerals be worked at the place covered by the application being inquired into, or somewhere else?' (Stevens, 1976). This criticism can be seen in a broader context, that is, the need, by default, in certain inquiries to examine the strategic policy context for a specific application. It has been argued that, 'if national policy considerations are to be excluded from the remit of public inquiries then the public must have confidence in the manner in which national policy is formulated and kept under review' (Rowan-Robinson, 1980). The need, in some cases, to consider national policy at a public inquiry is quite clear. At the Vale of Belvoir inquiry central government listed a number of broad issues which had to be considered. Although it may be better that some form of strategic policy should emerge from an inquiry, it is an admission that planning authorities at national, regional and local levels have failed to ensure that the future development of minerals is based upon sound strategic principles.

An alternative form of inquiry which, by definition, is charged with the responsibility of examining strategic policy, is the planning inquiry commission. A special inquiry of this type would be able to consider national and regional policy, to sponsor independent research, to examine the technical and scientific aspects of an application and to investigate a range of alternative sites. No planning inquiry commission has yet been set up, although a two part inquiry has been used to consider the National Coal Board's application to mine coal in the Vale of Belvoir. In September 1978, the then Secretary of State for the Environment, argued against using a planning inquiry commission to examine the Vale of Belvoir application on the grounds that other public inquiries had considered matters of national policy and that the first stage of a planning inquiry commission might prejudice the more formal second stage. In the event a two stage inquiry went ahead; a preliminary meeting was held to establish basic facts and arrange an agenda, the formal inquiry followed. It has been noted that the chief advantage of a planning inquiry commission, its ability to sponsor independent research, was lost and that the objectors were largely dependent upon the National Coal Board for technical information (Winward, 1980).

The failure to provide strategic guidance can lead to the consideration of an application, or local plan, at an inquiry. Although, in the case of minerals, there are shortcomings in the inquiry system a total remedy is not to be found in either, wholesale modifications to the rules which govern inquiries, or in the adoption of a planning inquiry commission. What has to be considered is the reason why matters concerned with minerals planning are subject to public inquiry. It has been argued here that, in part, the reason for such occurrences is the inadequacy of much strategic policy and planning at national, regional and local levels. The Stevens Committee considered this matter and concluded that, for example, the establishment of county mineral consultative committees would provide a forum for the discussion of long term strategic planning and the consideration of alternatives. Should such committees be established, then it would be reasonable to expect that the number of cases considered at inquiry would be reduced and that the 'ideology of public participation' described by McAuslan (1980) might be better served.

LEGISLATIVE ADJUSTMENTS

The purpose of this part of the Chapter is to give consideration to current attempts by central government to provide more adequate direction to planning authorities when dealing with minerals issues. It focuses attention on four areas; government statements affirming the county as the competent authority to deal with minerals, government guidance to Regional Aggregate Working Parties, revisions being considered to the Green Book, The Control of Mineral Working and the Minerals Bill currently before Parliament.

Minerals planning has been regarded for a number of years as a matter having strategic significance and as such appropriate to be dealt with by county planning authorities. This was recently reaffirmed by the Department of the Environment (Circular 2/81). Although it is difficult to understand why certain matters are deemed to be of sufficient strategic importance that they have to be dealt with by a county whilst others are not, it is encouraging to see that minerals have remained as a county matter . Even so, there can be little doubt that there are relatively few powers available to county authorities which allow them to formulate realistic plans for minerals which are based upon a wider consideration of regional and national needs and thus in the context of this text the question must be asked, to what extent does this represent a commitment by government to strategic planning or does it remain simply a matter of administrative expediency.

It is only fair to restate the point made earlier in this Chapter that there have been attempts, at regional, subregional and county levels, to put key issues such as minerals into a strategic context. However, for reasons already outlined, this has not always happened. Structure planning and some of the subsequent subject or local plans have attempted to provide such a strategic context for minerals planning but have found it difficult to accomplish anything worthwhile in the absence of national and regional strategies and related planning guidelines and policies.

Under the 1971 Town and Country Planning Act structure planning authorities were required to take account of related matters in adjacent administrative areas. How effective was any such inter county dialogue on matters of common interest during the preparation of structure plans could undoubtedly form the basis for a major investigation in its own right but even where regional working parties have been established to look at minerals nothing has yet emerged which could be described as providing an appropriate regional context for the review of structure plan policies or the preparation of subject or local plans. Albeit with the benefit of hindsight, the authors feel that the 1971 Act should have advocated the setting up of regional working parties or intraregional strategy groups for the consideration of minerals and other issues of strategic importance.

Government attitudes would seem to suggest that there is some appreciation of the fact that minerals is an area appropriate to strategic planning. This view would seem to be substantiated both by the retention of minerals as a county matter and by attempts to encourage counties to meet together to discuss minerals issues, particularly aggregates. However, there appears to be equal evidence to suggest that government currently lacks the will to provide an effective context for the strategic planning of mineral resources. The abolition of Economic Planning Councils and the absence of any new regional strategies leaves aggregate minerals as one of the few matters for which planning is being attempted at a regional level.

As far as aggregate minerals are concerned it is possible that government feels that it has done sufficient by encouraging the establishment of regional aggregate working parties in that these minerals have been supplied historically on an intra regional basis. However, for well over a decade it has become increasingly apparent that there has been an increase in the inter regional movement of minerals and that forecasts were starting to indicate that despite rising transport costs, if other policies such as conservation of agricultural land and Areas of Outstanding Natural Beauty were to be upheld, then certain regions would have to become net importers of aggregate minerals. Equally, the attitudes of successive governments during the past decade towards other minerals appear to have lacked a firm commitment to the establishment of explicit strategic planning policies, witness the random granting of exploration licences in the early 1970's and the apparently 'free for all' approach to onshore developments associated with the offshore hydrocarbon industry.

During the past twenty years there has been no up dating of the Green Book, The Control of Mineral Working. This has provided the only comprehensive attempt to guide local authorities when dealing with minerals planning applications. Clearly there have been changes in planning legislation since 1960. The Green Book does refer to 'national need' but at no juncture does this appear to have been adequately expanded upon by government except in the case of coal and the other energy minerals. It would be wrong to construe the Green Book as containing much of relevance to strategic planning. It is a document which was prepared to help with the processing of minerals applications and the preparation of some local plans. There is no doubt that the document has provided useful guidance over the years and that any revised version that may appear will be of equal value to local authority planners. It is worth noting here the greater success in providing strategic guidance in Scotland where national guidelines have been made available to regional and district authorities (Scottish Development Department, 1974 and 1977).

It is understood by the present authors that the proposed revision to the Green Book is more an attempt to bring the existing document up to date than to establish a strategic framework in which national and regional policies could be formulated to provide the essential context for minerals planning at a local level. As a guide to local authority planners the Green Book does have a role to play, but there is certainly a need for government to find an effective way of developing a machinery whereby minerals can first be considered at both a national and regional level.

It is of interest to note and particularly relevant to the stance taken within this text that, in a document giving guidance to aggregate working parties, circulated by the Department of the Environment (1979), two important points were made. In the first place there was a reaffirmation that minerals should and would lie within the control of county planning authorities. Secondly it was suggested that there was a need to have a wider regional and national context for local planning in order to avoid the somewhat arbitrary nature of current decision making. This view was developed out of the findings of the Stevens and Verney Committees. The five points which the Department of the Environment (1979) put forward as a strong argument for developing guidelines were:

(i)   provide a clear framework within which local planning authorities could develop mineral policies in Structure and Local Plans;

(ii)   serve as a consistent national framework for the Secretaries of State in consideration of the aggregates element of Structure Plans and the individual cases which come before him;

(iii)   help reduce the number of cases going to appeal;

(iv)   maintain a steady and adequate supply of material to the construction industry at minimum social, economic and environmental cost, and

(v)   enable informed debate at a national and regional level on the implications of current policies for aggregates.

It was seen that the basis for arriving at acceptable recommendations for policy and ultimately for action should be achieved through co-operation and collaboration between local planning authorities, the government (Department of the Environment and Welsh Office) and the minerals industry.

Having set regional aggregate working parties in motion it was clear this would prove to be a useful arrangement for the collection and processing of data and for encouraging a fuller debate within regions concerning matters of supply and demand.   The need to have reasonable consistent data collected in all regions was appreciated and to this end the Department of the Environment (1979) suggested a number of headings under which data should be collected.   These included inter regional flows, capability to maintain supply and the possible use of alternatives.   It is heartening to see such a document eminating from a government department and to see new ideas emerging.   The document outlines ideas for intensifying the activities of a National Co-ordinating group, established to discuss the results of the early reports which were produced by the various regional working parties.   In May 1980, ten months after the original paper was circulated, another document was produced highlighting the need for the working parties to be mindful of the implications of a sudden up turn in demand for aggregate materials as the country emerged from recession, thus giving credence to the present authors' earlier stated view that any strategies for minerals should be robust and capable of a measure of flexibility.

Although the Stevens Report did not favour the adoption of a national plan for all minerals, it is difficult to know how a greater degree of strategic thinking can be injected into minerals planning in the absence of a broader basis of policy for the development of the nation's mineral resources.   To date little progress has been made on the suggestion made by the Stevens Committee of developing these necessary policies for individual minerals though the approach proposed by the Department of the Environment (1979 and 1980) for aggregates has much to commend it for all minerals.

Minerals issues were not included within the 1980 Local Government, Planning and Land Act although they were included in some of the earlier drafts of the Bill.   Eventually minerals issues were presented in another Bill which has not become law at the time of writing.   Given the importance that was attached to the work of both the Stevens and Verney Committees it might reasonably have been assumed that following the decision to produce separate minerals planning legislation the new Minerals Bill would have referred to strategic as well as site development issues.

There is nothing explicit in the Minerals Bill which gives any support to the concept of preparing a broader strategic planning framework which could guide and influence the work of minerals planning at a local level. Despite the inclusion of some new ideas, it is essentially a document which amends and enforces existing local planning approaches to minerals. Commentaries on the Bill would seem to reinforce this view. For example, Eve (1981) identifies four main points of the Bill,

1. To improve the conditions that can be imposed on a Planning Consent for the restoration of mineral working. In particular to ensure that land that is to be restored to agriculture, forestry or amenity uses can be managed in a suitable way for a period after the restoration itself is complete to ensure that it returns to a proper level of productivity.

2. To provide that existing conditions on a Mineral Planning Permission can be changed with reduced or no compensation payable provided the changes do not affect certain fundamental elements of the Permission.

3. To deal with mineral workings where the working itself has ceased but the Planning Consent still remains outstanding. Two new Orders can be promoted; a Suspension Order, where the minerals are being held for future working and a Prohibition Order where the Operator has no intention of working any minerals that remain. Both Orders allow a certain number of new conditions to be imposed, again with reduced compensation being payable.

4. To put Mineral Planning Authorities, (County Councils except within the GLC area) under a duty to review the planning situation in respect of every mineral operation within their area. This review is to be repeated as frequently as the County Council determine. The carrying out of the review, however, is not a prerequisite to the use of the new powers contained elsewhere in the Bill.

From these points it would seem evident that no new strategic initiative is proposed.

Government's justification for adopting this approach in the Minerals Bill may be linked to structure planning. The structure plan process is still in operation, it represents a strategic planning approach in its own right and contains opportunities for monitoring and review. It could be argued that an approved structure plan provides an adequate strategic context for the detailed consideration of any development proposals including those to extract minerals. A further explanation of the approach adopted in the Minerals Bill might be that the Stevens Committee rejected the idea of a national plan for minerals. However, this is speculation and the final section of this text considers ways in which the Minerals Bill might be improved upon.

It would be unfair to deny the absence of any strategic planning potential within the Bill. The fact that minerals is retained as a county matter and not a matter for district authorities is not without some significance. Equally, it addresses itself to the question of review powers for existing permissions (see for example Jelley, 1981) which would be used as part of the mechanism for implementing a strategic policy for particular minerals. At a later stage in the Bill consideration is given to matters pertaining to aftercare treatment which might well be relevant to resolving broader land use issues such as the conflict between mineral extraction and agriculture.

Although there is some strategic potential contained within the Minerals Bill, its main thrust is directed at controlling mineral developments at a local level. It should be made clear at this juncture that the present authors are not denying the need for effective local planning control over mineral workings, giving due regard to a range of issues including, ownership, compensation, highways, rights of way and operational and aftercare considerations. Rather, what is being argued is that without an effective national and regional strategy for the development of minerals local action or control can become random and arbitrary.

9 Testing Evaluation and Assessment

During the past ten years the attention of planners has increasingly focused upon the difficulties involved in the assessment of major projects.   Both the Stevens (1976) and Verney (1976) Committees discussed matters of assessment at some length and, in addition, certain innovations in assessment methods and procedures have recently been introduced.   Many of these innovations relate to mineral development; the extraction of offshore oil and gas, the extension of coalmining and the advent of the super quarry have all required detailed evaluation and assessment.   The credibility of traditional forms of assessment used for many years by planners and claimed to provide comprehensive assessment have been brought into question.   Doubts have centred on traditional modes of assessment which have as their focus estimation of the impact and effects of a single activity at a specific location.   There are an increasing number of projects where the impacts of development are likely to be 'large scale and complex' (Catlow and Thirlwall, 1976).   These larger projects often span a series of sites or could be located in a variety of places.   By way of example it is possible to cite the recent applications for planning permission submitted by the London Brick Company which are currently the subject of consideration, or the wide range of coastal locations which have been considered for use by offshore operators.   Other projects are site specific and in these cases a major issue that confronts planners is the consideration of need, often within the context of national and regional patterns of supply and demand.   In a strict sense planning authorities can only consider those matters which relate to the submitted application.   However, from the records of many public inquiries it is clear that wider policy matters have had an influence upon decisions, for example at the Vale of Belvoir Inquiry.

Certain basic principles have been established by central government for the assessment of applications for mineral working.   These were laid down in both issues of the Control of Mineral Working (Ministry of Town and Country Planning, 1951 and Ministry of Housing and Local Government, 1960), the Grey Book and the more familiar Green Book.   The five main issues identified in these advisory booklets as forming the basis for planning control deal with both strategic and site development matters.   They are listed in Chapter 7.   More recently the Scottish Development Department in their National Planning Guideline on Aggregate Working (1977) have argued that in assessing applications for mineral working local planning authorities should evaluate and consider 'whether or not development is acceptable depends on a number of factors, including the scale of intrusion, the surrounding topography and the presence or lack of any artificial landscaping or screening .... There may also be a choice in meeting transport or market requirements and in balancing the national economic benefits in relation to special environmental factors'.

A number of major issues are given prominence in the Green Book and the Scottish Guidelines; first, the matter of national interest (which has been considered in detail in Chapter 6), secondly, the relationship between mineral working and other land uses, thirdly, the environmental and social effects of mining operations upon areas, and fourthly, the question of operating conditions, after-use and restoration. These matters are given more detailed consideration in the latter part of this chapter.

However, before examining assessment and evaluation techniques and methods, it is necessary to look at the context for decision making and to relate this to questions of equity. Murphy (1979) has criticised the present approach adopted by British planning authorities when considering applications for large scale developments. He has argued that the consideration of a planning application normally takes place within a 'national and regional policy vacuum', and thus 'decisions are commonly made which are dominated by relatively local considerations'. In such a situation the consideration of proposals for mineral extraction will inevitably result in a decision which is not entirely satisfactory to all of the parties which are involved. The consequence of the confusion which surrounds the assessment of proposed mineral workings is reflected in the number of applications which are called in or considered directly by the Secretary of State for the Environment. One of the most worrying aspects of this procedure is that the 'general public is usually excluded from any consideration of a planning application relating to a large-scale project until the issue comes before a public inquiry'. (Murphy, 1979).

The confused and spatially inconsistent methods of assessment currently in use can in extreme cases lead to a lack of equity in the decision which is arrived at. Because many applications for mineral working involve a number of parties whose interests or objectives conflict it is necessary to ensure that 'consideration be given to the distribution of losses that are implicit in alternative planning decisions'. (Berry and Steiker, 1974). It has been argued that traditional methods of assessment have proved to be inadequate in resolving matters of justice and equity and that decision makers should pay attention to 'the losses borne by each of the interested parties rather than a summary index of society's net of gains over losses'. (Berry and Steiker, 1974). In part, the failures of the present United Kingdom approach to decision making can be seen to be the result of inadequate assessment and evaluation, but it can also be claimed that the procedural framework suffers from inherent weaknesses. It is difficult, and strictly speaking impossible, to consider alternative sources of supply, different sites, or other technologies at a public inquiry; whilst local authorities are obliged in the development control process to assess each application on it's own merits. The powers, established in the 1968 and 1971 Town and Country Planning Acts, to constitute a Planning Inquiry Commission have not been used. The assessment of applications and decision making thus relies, in the main, upon the capability and ability of individual local authorities.

Three major aspects of evaluation and assessment are considered here. First, a brief review is presented of the procedures used for the incorporation of mineral issues within the planning process. Attention is focused on the distinguishing features of mineral working and the attempts which have been made to provide strategic guidance. Secondly, partial techniques of assessment and evaluation are examined, these techniques are those which have been used by planners to attempt to resolve the problems encountered in dealing with mineral applications. Thirdly, a brief review is presented of recent suggestions for more comprehensive forms of assessment, attention is focused on the potential use of environmental impact assessment.

PLANNING PROCEDURES

General advice to local authorities on the assessment of applications for mineral working has been provided by central government, most notably in the now out-dated Green Book. This advice, when taken in parallel with the range of powers conferred upon local planning authorities under the various Town and Country Planning Acts, provides a framework for the procedures which are used. Mining operations are classified as development, so all mineral workings and associated activities, require planning permission. Given that the majority of mining operations occur over a long time span, then permissions are normally sought over a considerable period, although in some cases mining operations have been suspended for a number of years. The scale and rate of mineral extraction is also a subject of concern to local authorities; planning permission may be granted on the assumption that a given working will operate at a stated rate of production, frequently such initial estimates are exceeded or not achieved. In order to prevent such problems, conditions regarding the rate of extraction may be specified by a local authority at the time when planning permission is granted.

It is not intended here to elaborate in great detail on the development control procedures used by local authorities; rather, it is the authors' view that the Stevens (1976) recommendation for a 'special regime' provides the essential argument as to why mineral working can be distinguished from other development operations. It has been argued by Rodmell (1976) that the Stevens Committee convincingly distinguished mineral working from other development operations on the grounds that:

a) mining operations are an end in themselves,
b) mineral working is destructive,
c) minerals can only be worked where they exist,
d) it is inevitable with mineral workings that at some time in the future all the mineral will have been extracted and that use of the land must cease,
e) a mineral operation is continuous; conditions have to be framed to allow for this and the planning authority needs officers correspondingly more knowledgeable in the processes involved,

f) consideration has to be given to changes occurring in mineral extraction technology.

These arguments led the Stevens Committee to consider a new system of control outside the planning system on a regional basis.   However, in the end the Committee recommended that control should be exercised by county planning authorities, and that a 'special regime' should be established with a sub-class of planning permissions to be called mineral permissions.

The Stevens' recommendations, based upon clear arguments and supported by much detailed evidence, were not fully accepted by central government (Caisley, 1978).   The present position is, therefore, one where county planning authorities exercise control over mineral workings within the general planning system and where development control procedures vary from authority to authority.   Steps have been taken (for example, through the use of standardised special mineral extraction application forms) to ensure a higher degree of regional and sub-regional coordination, but the essential truth remains that a county's ability to deal with minerals is as good as the officers that it employs.   That standard of ability is generally high; yet there remain instances where the criticism levelled by the Stevens Committee still holds true, that 'many of the past failures of planning control in relation to mineral workings were directly attributable to a lack of necessary professional skills in local planning staffs'.   (Stevens,1976). This absence of the necessary specialist personnel has been exacerbated by recent financial and staffing constraints imposed upon local authorities.

The consideration of applications for mineral working by local authorities is frequently time consuming, each application will vary considerably even if it is for the working of similar deposits.   The problem that faces local authorities is twofold; firstly, how to allow operators to proceed with the most economic form of mineral extraction, and secondly, how to ensure that the evaluation and assessment of an application takes account of other policies and objectives relating to land use, economic, social and environmental considerations. Although these issues have to be resolved in all cases, the potential for divergence between the various arguments is more significant in the operation of large quarries. When faced by an application for mineral working the local authority is therefore in a difficult position, put simply, 'in the case of a vital mineral supply the adoption of the policy that the polluter must pay obscures the fact that, since the commodity is vital, the consumer, that is society as a whole, must pay' (Down, Stocks and Pryor, 1976).   To the extent that most mineral operators claim that the mineral, which is the subject of their application, is vital, then it is correct to assume that local authority procedures will encompass a judgement of the balance between the conflicting requirements of the parties involved.

In order to attempt to resolve the problems that face them in making development control decisions, some county planning authorities have prepared minerals subject or local plans. These plans are intended to provide a link between structure plan policies and development control. Although set within structure plans they are far more detailed and elaborate. A local plan normally consists of a map and a written statement but can contain other material to illustrate or explain the proposals in the plan. Normally, the plan's proposals will conform generally to all structure plan policies. Until recently local and subject plans could not achieve official status until the structure plan had been approved. (South Yorkshire County Council, 1978). The form and content of local subject plans will vary according to the range of problems encountered within the county area and the level of detail in which mineral issues have been expressed in draft or approved structure plans. In the majority of cases the local plan time horizon will conform to that for the structure plan, however permitted reserves will be considered over a longer period. Local subject plans provide an important framework for the assessment and evaluation of individual proposals, they enable scarce staff resources to be concentrated on a particular topic, they are capable of revision more rapidly than structure plans and can provide detailed guidance for development control. There are, as yet, few local subject plans in existence, as such their value in practice has still to be tested. Two final issues must be borne in mind regarding such plans; firstly, that revisions to local subject plans may cause conflicts to emerge vis a vis other structure plan policies and secondly, that the detail provided in such plans may cause an undesirable concentration on site specific rather than strategic matters.

In the interim, before local or subject plans are available, strategic guidance for development control is provided through structure plans. The general form and function of structure plans has been outlined in Chapter 8 of this text. The majority of published structure plans contain a section concerned with the major issues related to the extraction of minerals. The level of detail in structure plan policy statements varies from county to county, depending upon the strategic significance of minerals within the plan area and the level of detailed knowledge held by the authority. As has been noted above, increasingly county planning authorities provide broad guidance in their structure plans and then elaborate and detail those policies through the preparation of local and subject plans. In certain structure plans specific issues are given more detailed consideration, for example, where a major new mineral working is proposed or where there are existing problems associated with the extraction of a mineral. The West Midlands County Council's 1980 Structure Plan includes a consideration of the likely consequences of any new colliery development in the southern part of the Warwickshire Coalfield. An assessment is incorporated within the plan of the consequences of mining activities upon the allocation of land for both industrial and residential purposes. (West Midlands County Council, 1980).

Other structure plans have given great emphasis to the problems associated with the extraction of minerals and the effects of mineral working upon other land uses and policy areas. This intermeshing of mineral policies with other matters can be clearly seen in the Peak District National Park Structure Plan of 1976. Here, the relationship between mineral working and transport, employment and recreation policies is starkly demonstrated. The scale of the problems faced by the Peak Park Planning Board led them to propose severe restrictions on any new mineral workings, whereby 'only a limited amount of mineral working is likely to be permitted in the National Park'. (Peak Park Planning Board, 1976). This strategic guidance was subsequently modified by the Secretary of State for the Environment who, in his modifications, suggested that 'proposals for minerals exploration will be considered on their merits'. (Department of the Environment, 1978).

It is clear from this brief review of the procedures which are used for the assessment of planning applications, and the importance of minerals issues in the preparation of guidelines, that the development of mineral resources presents a number of specific and difficult problems. Firstly, mineral working can be distinguished from other forms of development in that it is a destructive process which can only be undertaken where a resource exists. Secondly, a different and complex range of skills and experience is required of planning officers when dealing with minerals issues. Thirdly, it is often difficult to fully incorporate in structure and other statutory plans the specific issues that arise from mineral working. Finally, the extraction of minerals has a strong interaction with, and influence upon, nearly all other aspects of planning policy. There still remains a strong case for the operation of a special regime for minerals set within the context of strategic guidance prepared at both regional and county levels.

PARTIAL ASSESSMENT

As has already been stated certain general principles for the assessment of applications for mineral working were established in the Green Book of 1960 (Ministry of Housing and Local Government, 1960). Notwithstanding the existence of these general principles the extraction of minerals presents a number of problems in evaluating the environmental and social effects of mining. The Verney Committee (1976) stated that the consideration of an application for mineral working would, 'involve assessment of numerous and complex factors'. The range and complexity of the matters associated with mineral extraction has made the assessment of site specific planning applications very difficult, necessitating the employment of specialist staff and consultants, and the use of specific techniques.

A distinction should be made at this point between testing and evaluation. Testing normally operates through a set of criteria, for example, standards for the control of noise or dust exist and are applied. The use of such criteria enables a planning authority to decide if a particular application meets the minimum requirements laid down for that operation. There is little doubt that

the majority of local planning authorities are fully capable of implementing
such testing procedures, furthermore, specific legislation and codes of practice
exist which define the standards to be achieved.   The tests which are applied
by planning authorities are wide ranging and they include a recognition of the
operational problems that face firms.   It is important to stress that some of the
criteria, certain of which may be enshrined in the conditions which are
attached to a planning permission, are also used for the purposes of monitoring.
As such they provide a continuous basis for the assessment of workings and the
review of any conditions which may have been imposed.   In acquiring information
about a particular proposal for mineral working local authorities require
applicants to complete a standardised application form, this form is far more
detailed than that which is normally used by planning authorities and, amongst
other matters, asks for details of the methods of working and processing minerals.

Utilising the information provided by the applicant, and applying the criteria
specified in national and local guidance notes, local authorities test a proposal
with regard to factors such as the scale and rate of working, the direction of
extraction, the incidence and amount of noise, dust and blasting and the routing
and frequency of traffic.   These tests are applied both to the mineral working and
the associated ancillary processes.   At this stage a number of other initial tests
are undertaken.   Questions of visual intrusion, (of the working and of any waste
tips) waste disposal, the effects on surface and subsurface water, the likely
incidence of subsidence and the intentions of the operator regarding restoration
are all considered.   These testing procedures form the basis for a full evaluation
of the proposal.

The process of evaluation brings together the results of the initial tests and,
in addition, relates the particular proposal to other activities within the area
and to the established policies of the planning authority.   To paraphrase
Lichfield (1970) the purpose of evaluation is to aid the planner during the
planning process and to assist the decision makers.   Evaluation is properly
undertaken when all relevant tests have been applied and all interested parties
have been consulted.   It is necessary during the evaluation process to widen the
scope of investigation; some of the tests may be repeated, but within a broader
context.   Thus, for example, the effects of a proposal on other activities and
land uses should be assessed.   Some effects are site specific, but have
significance within a region or nationally; this is clearly the case where a
proposed mineral working is located at a site of archaeological importance
(Griffiths, 1980).   In addition matters such as the scale of workings, visual
intrusion, subsidence and the effect of the working upon infrastructure, housing
and other land uses should be evaluated.   For some minerals it is also necessary
to assess the extent to which mining operations can provide a stimulus to the local
economy or alternatively, the degree to which mining pre empts other, future,
planning decisions.

During both the testing and evaluation stages detailed consultations are normally undertaken with a range of government statutory, public and private organisations. Preliminary testing requires information to be provided by, for example, regional water authorities regarding the effects of the proposed mining operation upon surface and subsurface water supplies, upon drainage and also on the water supply requirements of the activity. In evaluating an application for development a local planning authority has to take account of the effects of a proposal upon other activities. It is therefore necessary to enter into further discussions with a range of consultees; in some cases these discussions and representations have a subsequent effect upon future minerals planning. This was clearly the case in the evaluation of the Selby coalfield, where due to the effects of mining operations it was considered that the main east coast railway line would be adversely affected by subsidence and would therefore have to be diverted; the construction work entailed by the diversion will involve the use of almost three million tonnes of aggregate. (Machin, 1980).

Methods of evaluation are numerous. They do nevertheless have one common feature in that they attempt to weigh the merits and failings of a proposal against each other in order to provide an informed basis upon which a judgement or decision can be made. Evaluations do not make decisions. However, it has been suggested that methods of evaluation should be used to generate recommendations that are technically defensible and politically realistic (Bate, 1980, b.) It is not always the case that such recommendations are acceptable to elected representatives. Other considerations, such as strategic requirements or political expediency, can result in a decision being reached which diverges from the initial recommendation made by a planning officer.

In the case of applications for small scale, non hazardous or short lived workings the evaluation of a proposal is normally a relatively simple matter whereby the application is judged against existing policies, designated land uses and the general amenity of an area. From this, and following a standard local methodology, the planning authority can develop a number of alternative recommendations. A final recommendation can thus be made which is in accord with the policies and guidelines specified by central government and the local planning authority. In many cases conditions are attached to a recommendation which, if accepted by the decision makers, form part of a consent. Such conditions will frequently be the result of discussions between the applicant and the planning authority, and will therefore have been considered as part of the evaluation exercise. The strength of any condition which is attached to a planning decision rests with the way in which it is phrased. Ambiguous conditions, which can later be subject to an interpretation far removed from the original intention, can lead to the initial evaluation being rendered useless. In the case of minerals applications, where the assessment of conditions is an integral part of the evaluation exercise, it is important that conditions are

realistic and enforceable.   Given the long life of many minerals permissions
then interpretations of conditions will change, a robust evaluation method should
encompass the likelihood of such change. (Tain,1980).

Evaluation is a more difficult matter where an application is submitted for a
large scale or potentially hazardous operation which may well cause significant
disruption or environmental damage.   In such cases planning authorities are faced
with the need to adopt wide ranging and more comprehensive methods of
evaluation.   Planning authorities have powers under Section 25 of the 1971
Town and Country Planning Act and Article 5 of the 1977 General Development
Order to obtain all the information that is necessary for them to be able to
consider an application.   These powers are used extensively.   However, in
some cases the authority has either failed to ask for the necessary information, or
found itself unable to examine the proposal in sufficient depth. (Catlow and
Thirlwall, 1976).   Welfare economics provided the foundation for most of the
evaluation techniques used by planners in the period since 1945.   Put simply the
aim of many such techniques is that the sum of those made better off should
exceed the sum of those made worse off (Roberts, 1974).   Planning balance sheets,
cost effectiveness analysis, financial appraisal and cost benefit analysis rely
heavily upon being able to assign costs and define effects in financial terms.
They are, by their very nature, partial techniques of evaluation in that they
seek to establish 'trade offs' and arrive at a minimum cost, maximum benefit
solution.   It has been argued that the use of techniques, such as cost benefit
analysis, in the assessment of planning applications has demonstrated their
inherent weaknesses (Clifford, 1975).

The weaknesses associated with partial assessment techniques are well
documented; they include problems of defining the need for a project, the
question of the range of alternatives to be considered, issues concerned with
measurement and comparability, the difficulty of assigning weights and the
matter of multiple sites.   Given the present legislative basis for planning,
it is sometimes difficult for an authority to question the need for a proposed
mineral working.   It is, however, a fundamental question and one which is
not answered satisfactorily by techniques of partial evaluation because many of
the factors which underlie the need for a project cannot be strictly quantified,
(see Chapter 6 of this text).   The range of alternatives that may be considered
in partial evaluation is defined by the authority undertaking the assessment.
In many cases this range is restricted and does not consider hybrid or mixed
solutions.   This failure to consider a satisfactory range of solutions is
attributable to incorrect usage rather than demonstrating any inherent
weaknesses in the techniques employed.   Measurement is a major problem; the
majority of the techniques require that the issues which are assessed are converted
into units of financial value and in many cases such measurements are
inappropriate.   The weighting of measurements has led to some bizarre

conclusions; it is difficult in partial evaluation to assign weights to effects that are almost impossible to quantify. The final problem is in the evaluation of multiple sites. Given that the majority of minerals planning applications relate to a single site, then the issue does not always arise. Where a number of locations are to be considered then the range of issues that have to be defined, measured and weighted will increase.

Techniques of partial evaluation cannot, therefore, provide planners and decision makers concerned with the evaluation of minerals applications with a simple or satisfactory set of answers or choices. To resort to such techniques can, if they are used in isolation, generate additional problems and difficulties. In essence, the techniques question the role of assessment within the planning process. It is the method of application of the techniques, rather than their conceptual structure, which has led to the search for more comprehensive methods of project assessment.

COMPREHENSIVE ASSESSMENT

Over the past 15 years planners, academics and other professionals involved with planning have attempted to develop better and more comprehensive methods of assessment. These attempts have been reflected in the efforts of central government to generate greater understanding of the potential role of environmental impact assessment, (Catlow and Thirlwall, 1976), and, more specifically, in the research which has been sponsored into the consequences of large scale mineral workings. (Down, Stocks and Pryor, 1976). In addition, the Commission of the European Communities is currently attempting to introduce a system of environmental impact assessment which will be common throughout the member states of the E.E.C. The European Communities' system will require a mandatory assessment to be prepared for all proposals for the extraction of solid fuels, bituminous shales, fissionable ores and metalliferous ores. Other mining proposals will be the subject of assessment if the government of the individual member state deems it to be desirable or necessary. (Commission of the European Communities, 1980).

Given this movement towards more comprehensive forms of assessment it is necessary to examine the role of such an approach within minerals planning. The Stevens Committee argued the case for a more comprehensive procedure for use in minerals planning; indeed, it is likely that had a 'special regime' been introduced then a form of environmental impact assessment would have resulted. In the absence of a 'special regime' it is possible that assessment procedures, designed to deal with a wide range of differing projects, apart from mineral working, will be applied to major proposals for mining and quarrying.

It is important to recognise that comprehensive assessments can be applied not only to individual projects, but also to policies, plans and programmes

(Lee and Wood, 1977). Broad policies, agreed at national level, which provide a context for plan preparation and the expenditure programmes of both the public and private sectors, can be assessed for their effects upon the environment. As can be seen from the operation of the Scottish National Planning Guidelines on Aggregate Working, (Scottish Development Department, 1977) it is possible to provide such a set of policy guidelines within which local or subject plans can be prepared, and which provide explicit direction for both regional aggregates working parties and local authorities. In the Scottish case it should therefore be far easier to incorporate the comprehensive assessment of an individual application within normal planning procedures.

It has been noted earlier in this Chapter that the extractive industries are notably different in character from the normal range of development operations. Such differences are crucial in deciding which procedure or method of environmental impact assessment should be used. Whereas, in the majority of developments, a distinction can normally be made between the impacts which are experienced at the construction and operational phases of an activity, in the case of deep mining it is important to add an afteruse phase, whilst for shallow quarrying a more useful distinction can be drawn between the operational phase and afteruse and reclamation. There are other important differences; the location of a mineral extraction is by definition fixed, whilst, the location of secondary and processing activities is not. Therefore, there is a case for a separate assessment of each of the constituent parts of a mineral extraction and processing operation. Furthermore many mining and quarrying operations create a number of external effects over a considerable period of time; important amongst such effects is the generation of additional road traffic. These external effects, because of their longevity, have to be considered as an integral part of any assessment.

Assessment procedures, methods and techniques should be designed so as to take account of the various types, forms and scales of both extraction and processing. In addition, they should be capable of distinguishing the impacts which are encountered during the construction, operational and afteruse phases of a mining activity. Of the various procedures which are at present available, the one suggested by the Project Assessment for Development Control (P.A.D.C) team would seem most appropriate to minerals applications. The P.A.D.C. approach (Clark, Chapman, Bisset and Wathern, 1976) provides a structured process for the consideration of major planning applications. One major innovation suggested by the P.A.D.C. team is that the prospective developer would be required to submit a Project Specification Report at the same time as the formal planning application. This additional report would provide detailed information on the nature of the proposal and allow for an initial consideration of the project to be undertaken. From this initial stage the local planning authority can then proceed with the normal range of consultations and, arguably more importantly, the construction and calibration of an Impact Matrix.   This Impact Matrix can subsequently be used to isolate the major effects and impacts in a Final Issues Report.   The submission of a final report and a set of recommendations to the decision makers, (normally the local planning authority) is thus based upon a structured and detailed assessment.   Two additional features of the P.A.D.C. method are noteworthy; firstly, stress is placed upon the need for discussions between the prospective developer and the planning authority and secondly, the impacts of the various phases of a project are the subject of detailed consideration and assessment.

The P.A.D.C. method would, with some modification, seem to meet the requirements of a local planning authority faced with the need to comprehensively assess a major proposal for mineral extraction.   What it does not do is to resolve one of the major difficulties that continues to bedevil minerals planning, that is, the fundamental question of the need to extract a given mineral.   In the absence of specific national and regional policies and plans for the extraction and processing of mineral resources this issue will remain unresolved.   Local and subject plans prepared by county planning authorities can, if comprehensively assessed, provide a strategic policy framework.   However, for certain major groups of minerals, especially the energy minerals, there is an urgent need for agreed national and regional guidelines if the arguments which have been rehearsed at numerous public inquiries are to be avoided. Cooperation between prospective developers and local planning authorities, as in the designation of mineral consultation areas, (Henry, 1980) might in part fill this policy vacuum.   Alternatively, regional level contingency planning can provide clear guidance to prospective developers (Grampian Regional Council, 1980).

What is clear from any examination of minerals in the United Kingdom is that there is a distinction between the best possible assessment that current practice can provide and the minimum level of assessment that is required by planning and other legislation.   With the increasing scale of mineral workings it is important to ensure that best practice becomes the norm.   Environmental impact assessment provides a procedure and a set of techniques that can ensure that planning authorities achieve this objective.   Both local authorities and commercial operators have found that the use of comprehensive assessment can accelerate the decision making process, (Royal Town Planning Institute, 1980) whilst the cost of producing an environmental impact assessment need not be excessive.   Whilst comprehensive assessment will not eradicate inevitable conflict, it can focus attention on the major issues and provide a structured forum for discussion and decision making.

SECTION IV

TOWARDS A COMPREHENSIVE APPROACH

10 Current Failures and Weaknesses

It is now thirty-five years since the establishment of the Waters Committee and almost ten years have elapsed since the appointment of the Verney and Stevens Committees.   In all three Committees, discussion was focused upon certain major problems that confronted and still confront planners.   The issues which have formed the focus of attention have changed in their relative importance but largely remain unresolved.   There remains a concern with avoiding any unnecessary disruption in the supply of minerals, whilst concurrently attempting to ensure that conflicts with alternative land uses are minimised.   The debate on the relative merits of economic growth and environmental conservation continues to tax the ingenuity of planners, government, and the minerals industry.   Matters of equity and of the social consequences of mineral working are still voiced, even in an era of low economic growth.   Planning practice, whilst often excellent, is still variable in quality and in any case is often forced to operate in a policy vacuum. These and other matters have not been fully resolved during the past three decades despite the efforts of government and the minerals industry.   Attempts to reach an acceptable set of solutions have, in part, foundered because of changing economic circumstances, but they have also been prevented from achieving success by a number of more general difficulties which have beset British strategic planning in the period since 1945.

The difficulties that have been noted throughout this text are summarised in this Chapter.   Although there are many praiseworthy features of British minerals planning, this Chapter is concerned with isolating the deficiencies that have been demonstrated in the recent past and which are likely to be perpetuated unless changes are made.   It is often easy to find fault; the authors are aware that criticism which is not constructive can lead to retrenchment and inertia.   The final Chapters therefore attempt to put forward some suggestions for modifications that might be made to the present methods and system of minerals planning.

ECONOMIC INEFFICIENCIES

It is difficult to isolate those problems of minerals planning which are solely economic.   The efficient working of a mineral from the operators' viewpoint can produce high or unacceptable social or environmental costs.   There are however, certain economic inefficiencies which have been generally recognised. Firstly, there is the question of the movement of minerals ; the transfer of materials between regions is not always necessary but nevertheless occurs . Secondly, there is still a reluctance to make full use of alternative materials. Thirdly, in a period of rising unemployment and low economic growth it is vital to maximise the multiplier effects of any mining development.   Fourthly, there is a need to more clearly identify the relationship that exists between the private and public costs which are incurred, either during the extraction of minerals, or following the cessation of working.   Finally, there remains the question of the efficiency of the planning system itself in relation both to the needs of operators and of society in general.

As has been demonstrated elsewhere in this text minerals are often hauled across considerable distances. Whilst it is inevitable that non ubiquitous minerals have to be transported from producing to consuming centres, it is often difficult to justify the movement of ubiquitous minerals across regional boundaries. The pattern of distribution of aggregate materials is overwhelmingly by road, yet the movement of such materials more than 25 km. to 30 km. is generally accepted to double their point of production cost. In 1977, 97·9 per cent of sand and gravel was handled by road, 1·8 per cent by rail and 0·3 per cent by water (Regional Aggregate Working Parties, 1980). With rising fuel costs it is likely that the delivered real cost of aggregates will increase in future. On this question it is worth noting that in fact the cost of aggregates as a percentage of total construction costs has fallen in recent years. However, this falling cost does not fully reflect the total cost of extracting or transporting materials. Long road hauls do create social and environmental disturbance and damage, whilst the use of heavier road vehicles, in an attempt to reduce the number of trips or haulage costs, can result in an increased demand for aggregates. If it is wished to reduce the consumption of aggregates, then by reducing axle weights a proportion of the 40 million tonnes of materials used for road maintenance could be saved. An increase in axle weight from 10 to 12 tons is estimated to double the damage done to a road surface.

The inefficiencies of the long distance movement of relatively ubiquitous minerals are reflected in other ways. It is likely that operators, given freedom of action, will wish to continue to make use of their own resources wherever possible. Yet this freedom to operate can create inertia which in turn may result in longer term supply problems. The greater understanding of the movement of aggregates that has been provided by the work of regional aggregates working parties has, as yet, failed to indicate the detailed reasons for long distance haulage, presumably in order to maintain commercial confidentiality. Should the demand for aggregates increase dramatically then regions, such as the South East, could face severe problems of ensuring an adequate supply of materials (Orrell-Jones, 1978).

If the long distance movement of minerals is to continue then the present attempts to establish rail depots, and perhaps super quarries, will have to be accelerated. In addition it may well be worth extending the use of other forms of transportation, noteably pipelines, which are at present mainly restricted to the conveyance of non ubiquitous materials. Given the high capital cost of establishing new production plant, pipelines are at present used to move limestone slurry from the South East to the West Midlands in order to meet the needs of the cement industry. Planning policy will have to move further towards the consideration of aggregate minerals and their least cost transportation paths on the basis of service areas, rather than within formal administrative boundaries. This may well require adjustments to approved structure plan policies, and will certainly necessitate close co-operation with both the consumers and producers of aggregates.

In certain regions there will be a continued need to import ubiquitous minerals, whilst in most regions non ubiquitous minerals will have to be brought from elsewhere. This will require an emphasis to be placed on providing an efficient transport system, which has as its twin aims, cost minimisation, and the avoidance of unnecessary environmental and social disturbance. The type of strategic guidance which has been provided by the Scottish Development Department (1977) would do much to help to resolve the issue of the movement of minerals. The definition of zones within Scotland clearly indicates the factors that will have to be considered in the location of new mineral workings.

Any attempts to reduce the unnecessary movement of minerals should be considered in parallel with investigations of the potential provided by alternative materials. Alternatives can be considered as both new sources of supply and as substitutes for existing materials. In the case of energy minerals the successive oil crises of the 1970's have already stimulated the search for new sources of supply and investigations of the feasibility of using alternative fuels. This can be seen in the interest shown by the National Coal Board in the coal deposits that lie beneath the Vale of Belvoir, South Warwickshire and Oxfordshire. Fuel substitution is now encouraged by national government and the European Economic Community; the recent increase in the use of coal and nuclear fuels for the generation of electricity is designed to reduce the demand for oil. To a certain extent similar developments can be seen in the consumption of aggregates. For some years, marine dredged sand and gravel have made an important contribution to the supply of aggregates to the South East region, 19 per cent in 1977. In order to minimise the further extention of sand and gravel working on land and to reduce the costly importation of materials from other regions, the extraction of marine reserves could be increased. The present licensing system for marine dredging would require further amendment if central government wishes to encourage the extensive use of marine resources. In addition, restrictions upon the use of marine aggregates in concrete would have to be revised.

Whilst alternative sources can play an important role in meeting demand, it is also possible to make further reductions in the production of land won aggregates by using substitute materials. As has been already noted, the substitution of minerals is feasible. By using waste or recycled materials it is possible to achieve a reduction in the use of new aggregates and also to assist in the clearance of derelict or despoiled areas. In 1977 alternative materials represented 7·5 per cent of the total production of aggregates in England and Wales, whilst the estimated cummulative national demand to 1991 sees alternative materials providing 8·7 per cent of needs. The efficient use of alternative aggregate materials could be increased, thus allowing a greater use of wastes and reducing the pressure on the use of new materials. Obstacles to the use of alternatives include the rigorous standards which are presently specified for building materials, the distribution of sources of waste and the rating of tips which are reworked.

An extension of this argument involves a consideration of the use of low grade deposits of various minerals. Constant redefinition of what constitutes a viable resource has and is taking place. Currently, such adjustments are associated with metaliferous and energy minerals. Low grade ores, which were previously uneconomic to mine, are now extracted and the waste tips of previous eras provide worthwhile deposits (Tanzer, 1980). It is likely that with increasing resource shortages then alternative sources of supply will become more important.

The role of mineral working as a provider of economic growth is the subject of much renewed interest. As has been demonstrated earlier in this text, many regional economies were initially founded upon mining activities. With rising and high unemployment and low economic growth it is desirable to maximise the beneficial effects of any future mineral extraction. This requires two major forms of action; the first priority is to ensure that, wherever possible, supplies to a mining industry originate from United Kingdom manufacturers, this general priority may be reinforced to stress the role of regional suppliers. The second form of action is concerned with the control of extraction (through depletion rates) and the distribution of any minerals or mineral products which are produced. The first form of action has as its justification the high level of the potential multiplier which is associated with mineral development. As in the past, there is ample current evidence from Scotland to suggest that mineral extraction can have a significant effect on the level and patterns of regional and national demand (Lewis and McNicoll, 1978). Central government has taken action in the past to ensure that national and regional suppliers are provided with an opportunity to compete for contracts, notably through the establishment of the Offshore Supplies Office. This type of initiative could be extended in order to increase the multiplier effects.

A broader definition of the consequences, of new mining and quarrying for a regional economy needs to be used if the maximum economic potential of such activities is to be realised. It has been argued elsewhere by the present authors (Roberts and Shaw, 1980) that major new mineral resources should be exploited as the leading sector in an integrated economic programme. Thus, for example, offshore hydrocarbons can act as a stimulus for the economic regeneration of a lagging industrial economy; even when the original source of the stimulus mineral is exhausted the revitalised economy will remain. This is an approach which has been suggested elsewhere, Miernyk (1975) has argued that the

118

expansion of coal production in Appalachia would not only affect the regional economy through a direct multiplier of 1·9, but also could attract and stimulate other activities such as aluminium reduction and coal gasification.   Miernyk concludes that if the 'new prosperity' which coal resource development will bring is used to encourage other investment then future economic development will be much easier.   He furthermore argues that 'Appalachia's economic future beyond the coal age will depend upon the extent to which the Region's political and business leaders acted to broaden its economic base at a time when funds were available to do so'.

In order to increase the economic efficiency of mineral resource development there is a need for political action to ensure that economic criteria are used to judge applications for mineral workings.   The Stevens Committee (1976) argued the case for the inclusion of a section on the 'economic justification for the working' on a standard form for mineral application.   Further political action might be required to enforce the contract preference obligations placed upon nationalised industries opeprating in the areas which are assisted under United Kingdom Regional Policy.   Furthermore, as has been suggested by Spooner (1980) amongst others, there is a case for the establishment of development agencies, using funds generated from the revenues of mining operations, in the regions affected by mineral extraction.   Political action to ensure that the future developmental effects of mineral working are maximised is necessary, and in some cases likely, at local level.   Concern has been expressed, in the West Midlands, that should any new coal mining take place in South Warwickshire then the economic impetus provided by coal extraction should be linked to more general economic development efforts.   Certainly the reworking and processing of waste materials would provide employment in lagging coalfield regions, whilst the handling and working of marine aggregates would assist in the redevelopment of dockland areas.   In advanced industrial nations it is inefficient to regard minerals development as operating on the basis of a resource frontier mentality.

The control of depletion rates and the distribution of minerals and mineral products also required attention.   If a resource development is allowed to proceed at too early a stage, or if extraction occurs rapidly, then there is a danger that local and regional industry will have had insufficient time to equip the mining industry.   There is a case to be argued that the timing of an operation and the rate of depletion should be viewed in the context of national and regional policies for the particular mineral in question and for the economic development of a local area.   In order that local firms might achieve the status of approved suppliers to a specific mining company, and thus be able to tender for contracts on an equal footing, then it might be necessary to delay or restrict mineral extraction.   Equally, it might be considered desirable to delay a mining operation until an indigenous company can extract the deposit.   This would avoid one of the problems which has been associated with attempting to maximise the economic effects of North Sea oil development, that is, the high degree of external control exercised over the oil extraction process. Had a slow, or slower, depletion rate been imposed upon the extraction of continental shelf hydrocarbons then the level of British participation in the North Sea might have been greater. This is not to argue a case for protectionism or the adoption of an island economy, rather it is based upon the view that in order to maximise the potential for economic growth provided by mineral wealth then central government should consider the ramifications of allowing early extraction to occur.

The processing and distribution of minerals and mineral products is an important source of potential added value. Although minerals may themselves be low cost or of low value, the processing of raw materials can greatly add value in relation to the weight or bulk of a refined ore and a finished product. This indicates, for example, the importance which is attached to the selection of locations for processing plants. Given the high level of capital investment required to establish processing and distribution facilities, then the multiplier effects of development are significant. Additionally, the indirect effects in terms of employment and the secondary multiplier can be important. Although in the case of modern large scale aggregate and stone workings the direct labour demand may be low but the economic impact can be considerable (Down, Stocks and Pryor, 1976). The same may well be true for other minerals such as oil, gas, potash and ironstone, but the direct labour demands of the coal industry are high. The cummulative manpower demand for the Selby Project has been estimated as reaching 3,972 by 1988 (Arnold and Cole, 1981). Capital investment in process plant depends, in the case of many minerals, upon the perceptions which are held by large multinational corporations of their international interests. Investment decisions can and do work against the best interests of regions and nations within which deposits of minerals occur. Whilst this is not a new phenomenon, indeed this was until recently the norm in the exploitation of mineral resources in less developed nations, it is an inappropriate and inefficient basis for minerals planning in advanced industrial societies. Controversy has occurred over the refining of British sector North Sea oil outside the United Kingdom, especially at a time when many refineries are operating below full capacity.

When so few opportunities present themselves upon which to base economic growth, it would therefore appear essential that mineral development is viewed not only as a matter for planning control, but also as a basis for economic revitalisation. In the absence of agreed national policies for all minerals, and at a time when many regional strategies and structure plans are becoming out of date, then an economic context for minerals planning is a necessity.

A fourth matter that has to be considered in examining the inefficiencies that exist in minerals planning is the divergence of public and private costs.

Inherent in any discussion of mineral resource development are questions of waste disposal, dereliction, loss of amenity and afteruse. Considered from an economic viewpoint these are important matters, in that, although the hedonic approach argues that there is no market for environmental quality, there are marketed commodities whose market value is affected by the environment (Pearce, Edwards and Harris, 1979). Thus, for example, house values in an area subject to mining could be affected by various forms of pollution. Even if doubt is cast upon the validity of the hedonic approach, then other economic measures can be used to estimate the effects of mining upon the environment. Some of these have been discussed in the preceeding Chapter; they include cost benefit analysis and environmental valuations based upon social surveys.

What can be observed, in planning law and practice, is that a value is placed upon environmental disruption. Compensation payments, afteruse conditions, restrictions upon working hours and upon operating methods are all attempts to redress any loss which is suffered by the granting of planning permission for mineral working. It is also important to note that other estimates or valuations have been placed upon the undesirable effects of mineral working. The Hunt Report (1969) observed that, 'While despoliation of the natural environment went unregarded when coalmining and other industries located on or close to the coalfield were booming, the dereliction which was left now imposes a significant economic penalty on the area around'. Although the Hunt Committee followed the work of many other authors and observers, it represents the formal recognition of the economic problems that can be associated with poorly planned mineral extraction.

Modern mineral extraction can generate a high level of demand for both economic and social infrastructure. It is accepted that much of such infrastructure is provided by the public sector, often entailing considerable expenditure. In order that the provision of infrastructure can be justified, then the expected level of returns from a mining activity should be clearly demonstrated at the outset. The development of offshore oil required the provision of many facilities in North East Scotland and the cost of such provision has to be set against any benefits which arise. This is not to suggest that improvements to economic and social infrastructure will not be of long term benefit to that region, but rather to suggest that firms who benefited directly might have made a greater contribution to the cost. Even accepting the argument that any increase in the extraction costs incurred by mining companies will, in the long run, be passed on to the consumer, nevertheless at least some convergence of private and public costs might be achieved.

Finally, there is the question of the economic efficiency of the current system of minerals planning vis a vis the needs of operators and of society in general. It is difficult to provide a comprehensive evaluation of this matter. What may seem to be an economic inefficiency in the planning system may well reflect political reality.   In broad terms there are four issues which require some attention.   Firstly, there is the fraught issue of the speed of decision making; secondly, the relationship between the consideration of an individual application and other strategic factors; thirdly, problems with regard to the type of evaluation which is used by planning authorities and finally, there is the matter of reconciling the local with the national interest.   One author has argued recently that minerals planning has little cause for optimism; Bate (1980, b) considers that the 'strictures of law, Government policy, geology and technology currently militate against desirable objectives'.   This need not be the case. Whilst the issues with which minerals planning is involved may be controversial, solutions can be found to many of the problems which are apparent.

Regarding the question of the time taken to reach decisions there are a number of points which have to be considered.   It is not always the case that local authorities are provided, by an applicant, with all the relevant information which is required before a decision can be reached.   Consultation prior to an application can considerably reduce the time which is required for a formal evaluation.   Allied to this, local authorities often have little by way of strategic guidance to which an application can be related.   Thus, in many cases an application has to be considered at inquiry.   Some local authorities also suffer from problems of understaffing, from a lack of basic survey information and from an overload of applications.   Additionally, there can be conflicts of interest within the authority, most significantly between officers and members (Bate, 1980, b).   There are clearly cases where unnecessary delay occurs but these would appear to be few in number despite the views expressed by certain of those who submitted evidence to the Stevens Committee.

The efficiency of the planning system would be enhanced if the type of strategic guidance provided by the Scottish Development Department (1977) was the norm.   The Confederation of British Industry has suggested that there is a need to move towards mineral planning on a national basis (Stevens, 1976).   A clearer definition of national and regional policies towards specific minerals would certainly be of great assistance to county authorities in the determination of applications.   If planning authorities are not provided with adequate strategic guidance, then in the case of difficult of controversial applications, delay is inevitable.   Likewise, public inquiries will continue to be subject to delay and to costly extensions of time, due to a lack of definition in national policy, and the virtual absence of regional planning.

Evaluation methods and procedures are not consistent between authorities. An applicant may be required, by one authority, to provide information which another authority may not request.   It is also difficult to link the activities of local authorities to those of national government (and by implication to national and regional policies) if there are wide variations in the approaches used for the evaluation of applications.   Standardised procedures and methods, based upon national guidelines, would allow for the consideration of applications to be expedited.

National and local interest will, even in an ideal situation, inevitably diverge. This matter has been considered at some length in Chapter six of this text. However, if there has been prior discussion about the role and importance of specific minerals, then robust strategic policies will have emerged. This may be a vague and somewhat pious hope, but remains the key to efficient minerals planning. Both the aspirations of society and rationality can be best combined through thorough and explicit strategic planning.

ENVIRONMENTAL AND SOCIAL CONSEQUENCES

The first part of this Chapter looked at some of the economic failings of the present system of minerals planning. Many of the matters that have been discussed have an environmental or social dimension. Other, more explicit, consideration is now given to the environmental and social consequences of mineral working, and the failure of the current planning system to fully appreciate and accommodate them. Attention is drawn firstly, to the difficulties of environmental assessment, secondly, to questions of equity and distribution and thirdly, to the long term social impact of mining and the problems faced by areas dependent upon minerals and other industries which are associated with mining.

The current debate within the planning profession and the minerals industry on the possible introduction of a system of environmental impact assessment has highlighted many of the problems which face the parties who are involved in mineral development. It is clear that it is possible to assign costs to many of the environmental consequences of mining, whilst it is also accepted that many of the non quantifiable effects of mining upon the environment can be the subject of agreed judgements. At one extreme it is possible to identify those areas which are so sensitive to change that they might be, as suggested by the Zuckerman Commission (1972), designated as 'protected areas'. In other localities the level of environmental impact that is experienced may be acceptable, although every effort should be made to minimise the long term consequences. This implies that planning authorities should in future ensure that any consideration of applications is based upon a detailed environmental baseline study of the area to be mined, that they should carry out a full impact assessment and then should continue to monitor mining activities at both the operational and afteruse phases. It has been argued (Murphy, 1979) that had the recommendations of the Dobry and Stevens Committees been fully adopted by government then such a procedure would have ensued. There are cases where the type of assessment which is detailed above have occurred, notably in the development of the Orkney and Shetland oil facilities.

Environmental assessment remains an urgent matter on the agenda for the debate of reforms to the system of minerals planning.   Despite the work of organisations such as the Royal School of Mines (Down, Stocks and Pryor,1976), Aberdeen University (Clark, Chapman, Bisset and Walthern, 1976) and Manchester University (Lee and Wood, 1977), there is little to suggest that the need for comprehensive environmental assessment is universally accepted.   Whilst not disagreeing with the observation of the Verney Committee that many environmental judgements 'must at present remain essentially pragmatic', there is a requirement that current decisions do not add to the already considerable amount of inherited dereliction.   As has been noted earlier, there is a large measure of agreement on the damage caused by despoliation.

The reaction of the minerals industry to the increasing pressure has been to intensify the level of concern which is demonstrated in submitting proposals for mineral extraction.   What is now needed is a procedure whereby this concern can be evaluated within the system of minerals planning.   The argument is quite simple; it is, that the view expressed recently, by the Deputy Director of the National Coal Board, should be incorporated within planning practice.   Moses (1980) states that, 'our case rests that we are a responsible body able to live in a spirit of good neighbourliness with others seeking to exploit the rural environment for commerce of pleasure'.   If this is accepted then the problem can be resolved.

Equity and the distribution of the costs and benefits of minerals development are difficult issues to incorporate within the planning system.   There is little in planning law to suggest that there is any need or requirement for local planning authorities to consider such matters when examining an application or preparing a plan.   What is apparent from the practice of minerals planning is that with a few exceptions, such as the Zetland County Act of 1974, there has been little effort made to consider the wider ramifications of mineral extraction or processing.

In order to ensure that matters of equity and distribution are considered as part of the planning and decision making process it is necessary for central government to take action.   Berry and Steiker (1974) have argued that the traditional approach, whereby disadvantaged persons are compensated, is not sufficient.   This is because the losers are not always fully compensated and, even if a payment is made, there are circumstances under which proper compensation is not possible.   It is in these exceptional circumstances, where losses are suffered which are irreplaceable, indivisible or not substitutable, that planners face particular difficulties.   The extraction of minerals does generate a number of situations where there are losses of this kind.   Under these circumstances, traditional methods for determining compensation may not always provide adequate or meaningful recompense to those who lose.   There are a number of ways in which minerals planning could attempt to improve the current situation. The adoption of environmental impact assessment can assist in identifying those who will lose and those who will gain should a mineral be extracted. If minerals planning operated within the context of clear strategic plans and guidance then the incidence of inequity could be anticipated and alternative sites or materials could be assessed. Furthermore, should it be difficult or impossible to resolve inequity at a local level, then at least it would be possible to examine matters of regional or national need and set them against local losses.

The distribution of any benefits which might accrue from mineral extraction is also broadly outside the remit of local authority planning. Despite the precedent of the Shetland attempt to distribute the gains obtained amongst a local community, there is little liklihood that such exercises will become commonplace. The suggestion made earlier in this Chapter of establishing a regional or local development fund from the profits of a mining enterprise could increase the extent to which benefits are distributed. At least it should be possible to ensure that local or regional communities do not face the situation whereby they provide all the overhead capital which is necessitated by a new mining project, but do not participate in any of the benefits. Furthermore, the adoption of the suggestion, made by the Zuckerman Commission (1972), for the establishment of a 'renewal trust' would assist in the rehabilitation of mining sites and thus prevent the costs of reclamation from falling upon the public purse.

With regard to the longer term social consequences of mining activities there is little to suggest that planning authorities can reasonably deny all responsibility. Some two hundred years of extensive, large scale, mining activity in the United Kingdom has had a severe social impact. The work of Rees (1978) and House and Knight (1967) provides a graphic illustration of the longer term effects of mining. On a shorter time scale, efforts have been made by the Peak Park Planning Board (1976) to minimise the impact of quarrying and mining upon villages. Other more explicit assessments of the short term social consequences of quarrying have been presented by representatives of areas which are subject to quarrying. It has been argued, by Fletcher (1980) and others, that in the case of, for example, sand quarrying which creates significant social disturbance then the market price of the produce should incorporate an element to represent social costs. This could then be paid in compensation to those who are adversely affected. Whilst such innovative solutions are generally outside the concern of planning control, other issues are not.

Modern mining and quarrying is far less labour intensive than early forms of extraction. Even so large new projects have relatively high labour demands and can thus generate irreversible social adjustments (Arnold and Cole, 1981). If it is wished to avoid creating a new series of 'D villages', that will haunt the twenty first century, then planning authorities must take action prior to the commencement of mining operations. It is accepted that the mobility of labour has increased and that the process of industrial transition has accelerated, nevertheless, the danger exists that lack of forethought will exacerbate the present situation of selective regional distress. Structure plans, local plans and minerals plans provide basic vehicles for ensuring that the errors of the past are not repeated. This is not to suggest that the potential benefits of extracting minerals should be foregone, rather that new mining activities should form the basis for regional economic and social development strategies. Indeed, it is the case that mining activities can, in certain areas, have an important role in diversifying employment opportunities as well as simply increasing the total number of jobs which are available.

It is, in the final analysis, the role of local planning authorities to ensure that the optimum conditions are obtained for the benefit of their resident populations. This presumably implies that whilst the economic opportunities of mineral extraction are to be encouraged the undesirable social consequences should be minimised.

THE STATE OF THE PLANNING SYSTEM

The failures and weaknesses of existing minerals planning have been indicated in relation to economic, environmental and social factors. There are a number of other important matters which have to be considered briefly, most of which are self evident from the preceeding discussion.

Firstly, there are the difficulties which confront local authorities when faced with a controversial application. These are due to the absence of explicit national policies or guidelines. This is a matter for central government to resolve; however, in moving towards the creation of a policy base, government has the support of many sections of the minerals industry (as seen in the evidence of, for example, British Industrial Sand Limited and the Confederation of British Industry, to the Stevens Committee).

Secondly, despite the generally high standard of minerals planning in local authorities, the scope and extent to which minerals issues are considered in structure plans is variable. Allied to this, not all county authorities have prepared, or are in the process of preparing, minerals subject or local plans. Thus, the coverage of mineral or mineral related issues will be partial. As there are no regional mineral strategies or plans in the process of preparation, (with the exception of the activities of the various regional aggregates working parties) then it is likely that the treatment of minerals issues will continue to vary between local authority areas.

Thirdly, the evaluation methods utilised by local authorities are not consistent, between authorities, or in the level of detail which is studied. The aim of all authorities should be to achieve the standard which is set by the best current practice. Staffing, data and other problems hinder many county authorities and although these are in part of their own making they are compounded by the attitude of central government towards public sector expenditure.

Finally, the inquiry system has, from the viewpoint of operators, the public at large and planning authorities, a number of crucial weaknesses. It operates, in the field of minerals, in a policy vacuum, it is often said to be responsible for delay and cost excalation and it cannot in the strict sense consider alternative sites or other means of meeting demand. No planning inquiry commission has yet been constituted; such a commission might help to reduce the difficulties encountered by so many public local inquiries.

In the final analysis minerals planning will only resolve the problems that face it if central government moves towards the adoption of a positive stance and provides clear guidance on policies for individual minerals and the operation of the local planning system. This would appear to be the major priority for reform if the minerals industry is to move forward profitably and with a measure of public agreement.

11 Alternative Approaches and New Policies

In the earlier Chapters of this text attention has been largely focussed on what has happened in terms of the organisation, management, planning and development of the nation's mineral resources. The aim of this Chapter is to look beyond current legislation, convention and practice in order to consider other possible ways of coping with the demand for minerals given the complex economic, social and environmental consequences of their development. Although some of the ideas developed within this Chapter might seem radical and perhaps unrealistic in the context of both planning practice and governmental and business attitudes, it is felt that they should be aired if they are to form any part of the wider debate about the improvement of minerals planning in the United Kingdom.

In the final Chapter of this book the authors do attempt to look at the way in which government and industry might realistically progress from the current situation to one where comprehensiveness and cohesiveness guide the formulation of policies for the planning and subsequent development of mineral resources within the United Kingdom.

METHODOLOGICAL ALTERNATIVES

One of the great difficulties that has accompanied many proposals to develop minerals is that of the conflict between present land users and proposed developments. The question of how planning machinery is to evaluate the merits of the cases put by those who propose and those who oppose mineral extraction is difficult to answer. It is made all the more difficult when the protagonists in the conflict are able to support their respective cases with explicit government policies pertinent to their own particular land use activities. There is surely a demonstrable need for a more explicit statement by government on the assignment of priorities in the formulation of policy. Simply, if new development proposals are likely to create conflict then a means of resolving it efficiently must be devised. Currently this issue is dealt with through the planning inquiry system but unfortunately this tends to consider applications only in a site specific context. It would be of interest to know how many appeals could have been avoided if governments had been more considered in the way in which their policies had been developed. It would of course be naive to presume that all conflict would be avoided by an edict from government, such as for example, 'minerals winning will always take precedence over agricultural and amenity considerations'. What is being suggested here is that governments should be moving towards a position of establishing integrated national and regional land use policies. These policies should be based upon a fuller understanding of the country's economic, social and environmental needs and be sufficiently robust to accommodate planned change over time.

The planning inquiry system is forced to take a blinkered approach and there has been continued resistance by successive governments to the idea of adopting

the broader based planning inquiry commission (Rowan-Robinson, 1980).
However desirable many might consider such an approach to be, and certainly
it was called for at the Windscale and Vale of Belvoir Inquiries, it is still
difficult to see how the full potential offered by the planning inquiry commission
could be realised in the absence of clearly and consistently formulated national
and regional land use, economic and social policies.

There are probably few informed practitioners, politicians and commentators
who do not accept that policies and plans have to be amended in the light of
changing regional, national and international circumstances.   As early as 1949
during the debate in the House of Commons on the National Parks Bill, Silkin
gave notice that in future, conservation and amenity considerations might be
expected to give way to industrial requirements.   This has been referred to
subsequently as the Silkin Test.   Although the Silkin Test appeared to indicate
government's acceptance of the need to establish priorities which could form
the basis for making unpopular and controversial land use planning decisions, it
could also be argued that it did nothing more than provide a loophole to be
exploited by political and industrial expediency.   What is clearly lacked was
firstly an explanation of why economic considerations should override those of
conservation and amenity and secondly, it did not provide a basis for deciding
at what stage and under what circumstances the policies of one government
department should fall victim to those of another.

Judging by the acrimony engendered at many public inquiries, there are those
who feel that the failure of the government to deal with the inconsistencies in
its policies is nothing more than a political ruse to avoid making a firm
commitment to planning policies and thus abdicating responsibility for providing
the essential national context for the formulation of regional strategies.   These
should provide the basis for local planning at county and district levels.   The
logic of the family of plans, implicit in the 1968 Town and Country Planning
Act, has never been given an opportunity to prove its worth as far as minerals
planning is concerned.   This is not likely to be the case until central
government becomes more explicit in identifying the key elements in its social,
economic policies and in establishing an effective machinery for coping with
the complexities which manifest themselves at an inter regional level.   It is at
the local level however, that most planning legislation and procedures have been
directed.

It will not be possible to carry out effective planning for minerals at a
regional level without more comprehensive information about the extent and
distribution of all such resources.   There would appear scope for government to
initiate a comprehensive survey of the nation's mineral wealth including those
deposits which are readily workable and those which for technical or economic
reasons are not viable at the present time.   It is true that in the early 1970's
the government issued exploration licences to companies and individuals to

prospect for minerals throughout the country (Blunden, 1975). Almost ten years later it is difficult to guage if anything of worth resulted from what were in the final analysis somewhat sporadic activities. A comprehensive survey of the nation's resource base could capitalise upon the expertise of existing organisations with a proven capability such as the Institute for Geological Sciences (Stevens, 1976). It remains a fact without an adequate and up to date data base effective national and regional policy making will not be feasible.

Analysis of data has formed something of a stumbling block to participants in planning inquiries. Different groups have a habit of arriving at different conclusions using the same basic data (Ley, 1976). Attempts at becoming objective in the analysis and interpretation of data through the development of cost benefit analysis met with considerable opposition at inquiries because of certain inabilities to quantify and cost some of the elements involved. To this end, effective means of comprehensive analysis must be sought, bearing in mind the need to highlight those disadvantages which partial techniques of analysis have shown themselves unable to identify and quantify (Clifford, 1975).

Industrialists and developers often argue to government that market forces alone should determine the use of land for development. Even though the most committed disciple of the principles of the market might accept a need for some intervention and control over land use activities, it does appear that some industrialists and developers have a blind spot when it comes to assigning the full range of adverse impacts which might result from their actions (Searle, 1975). Even on issues of costing their view does not seem to include the loss of production and income to other industrial sectors which may result. Sometimes it appears that little consideration is given to the social and environmental costs to society which might result from their activities. Although it would be foolish to ignore the influences of the market completely, certainly there would seem to be every need to give adequate thought to all the implications which eminate from a decision to permit development. There have been attempts to develop a total costing framework for the analysis of industrial decisions (Humble, 1973).

Consideration has already been given to the opportunities offered by environmental impact assessment as a method of analysing and assigning the full implications of particular developments. Without wishing to rehearse the complete range of possibilities offered, it is important to stress that government should be adopting a far more positive attitude to the inclusion of this technique within the range of evaluatory methodologies currently used in planning. This presupposes that government adopts fully the arguments for a common European system of environmental impact assessment. To date, despite some work commissioned by the Department of the Environment (for example, Catlow and Thirwall, 1976), successive governments have shown a marked reluctance to give wider support to its use within statutory planning.

It is not altogether clear why governments are reluctant to move towards the adoption of these impact methodologies. There is a feeling in some quarters that existing methods and procedures are adequate and that environmental impact analysis adds nothing of extra value (Thorburn, 1978). Equally there is a belief held by some that the experience of its use in the United States of America is that it has proven to be far too difficult and cumbersome to deal with and has perhaps added more problems than it has actually solved. Stoel and Scherr (1978) suggested that its use in America had brought about a more open form of government. Any system which required government and industry to divulge information which traditionally they might not have done, may be yet another reason why government and industry are not keen to see such procedures adopted in Britain. However, given current moves to introduce a mandatory system of environmental impact assessment within the European Economic Community for certain classes of development (Commission of the European Communities, 1980) it would seem not unreasonable to suggest that the British government should be preparing itself accordingly. Many important mining activities, especially for energy minerals and metaliferous ores, are considered as requiring a mandatory assessment whilst all other developments for the extraction of minerals are likely to be subject to assessment at the discretion of national governments.

The reliance on partial evaluation techniques has arguably been a major fault in the appraisal of major developments throughout the world. Needless to say that where such partial techniques have been used there has been a tendency to emphasise those aspects of the project proposals which can be readily quantified and to minimise, if not ignore, those which cannot. There seems to be a general reluctance to explore other methods of appraisal and forecasting. Whilst not wishing to divert attention from important opportunities offered by comprehensive modes of assessment there would appear to be some justification in pointing out the value of more discursive methods which are based on available data, informed opinion and experience. Although no government might wish to commit itself to policy formulation solely on the basis of such approaches, the contribution that 'Delphi' and 'Scenario Writing' techniques for example, can make to the process of strategy formulation should not be underestimated.

Adopting comprehensive methods of assessment will generate a wealth of data and opinion whether this is done for policies, programmes, projects or plans. The magnitude of the task of coping with a large mass of information should never be underestimated. In the United States of America the use of environmental impact assessment procedures has resulted in literally volumes of evidence being placed before the judiciary by both applicants and objectors. The Carter administration felt obliged to issue guidance through the Council for Environmental Quality to help keep the evidence resulting from comprehensive analysis within manageable proportions (Stoel and Scherr, 1978).

There are many advantages to be gained from having a planning system that can operate efficiently and speedily. However, this should not be achieved by rejecting comprehensive forms of assessment on the grounds that they generate too great a volume of evidence to be considered at a planning inquiry. Lee and Wood (1977) suggest that environmental impact assessment can be used in a way that is both efficient and capable of instilling a greater degree of consistency into the planning process. This could be done by extending environmental impact assessment to higher levels in the decision making process, namely policies plans and programmes, as well as applying it to individual projects. This would provide a broader policy context for the appraisal of individual projects and it is anticipated that this would also reduce the number and length of public inquiries.

Any new approach to data collection, analysis and appraisal is likely to create difficulties for those who become involved in its operation. Bueeaucratic and procedural problems should not be overlooked, nor those of institutional inertia to the process of change. To advocate the establishment of new bodies or organisations would be unrealistic and arguably unnecessary, particularly at a time when central government is trying to reduce rather than increase the level of government. It is true that in the United States of America the Council for Environmental Quality was established to guide and advise those involved with environmental impact assessment, but in the United Kingdom organisations already exist which are capable of undertaking this form of assessment themselves or appraising those submitted them (if this were to become a part of a formal planning procedure). In order to prevent either the inquiry procedure or the Department of the Environment from being compromised it is therefore suggested that environmental impact assessment preparation should be assigned to an independent body. This organisation would perform a 'lead agency' role, co-ordinating the views of the various bodies concerned with the application. However, before any detail can be worked as to who and under what circumstances comprehensive forms of assessment should be utilised government should first accept the need to provide a suitable national and regional policy context for minerals planning at a local level and secondly, accept both the spirit and the letter of any European Directive requiring comprehensive appraisal of projects to be carried out.

Following on from the suggestions made by Lee and Wood (1977) it would seem appropriate and in line with the broad philosophy of this text to advocate the adoption and application of broad based and comprehensive analysis in national policies, plans and programmes. If government refuses to move towards providing a national and regional context in which coherent and effective local planning decisions can be made, as advocated in the preceeding Chapters, there seems to be little alternative but to advocate the adoption of environmental impact assessment for specific projects. Whilst appreciating the contribution it can make in the evaluation of individual planning applications, the present authors feel that without this broader political context there is no guarantee that appropriate decisions will be taken which are consistent with strategic objectives.

It would seem appropriate here to once again draw attention to the points made by Murphy (1979) concerning the value that environmental impact assessment could be making already in minerals planning. In looking at the whole question of planning control over mineral working the Stevens Committee (1976) expressed the view that the great majority of mineral development should be subject to Dobry's Class B procedures. If the Dobry recommendations had been accepted then the extended time period for the consideration of major development planning applications would most certainly have applied to proposed mineral development. Within this longer time period it would have been possible for a more effective dialogue to have taken place between the applicant and the planning authority, a further point recommended by Stevens. Although Stevens proposed the use of a standard mineral application form rather than subjecting minerals to environmental impact assessment procedures, Murphy further points out that this form would in fact cover substantially the same ground that would be covered by environmental impact assessment. In order that an assessment might be complete it would be necessary to add sections on the social and economic impacts and the consideration of alternative sites and sources.

LEGISLATIVE ADJUSTMENTS

There is little contained within the current Minerals Bill to spark the imagination of regional and strategic planners. As suggested in Chapter eight the Minerals Bill will, if enacted be a tool for local control and only by implication will it be of any conceivable value for broader regional and strategic planning. As mentioned previously there may well be an explanation of why it was decided not to include a section within the Bill which sought to provide an effective strategic context for future minerals planning. It is perhaps worth giving some thought to the means by which the proposed legislation might be improved to meet this criticism.

What is contained within the Bill will be of value to planners working at the county level processing applications to develop minerals and undertaking periodic reviews of mineral working and permissions granted and as such there seems every reason to support the Minerals Bill as far as it goes. Presumably the detailed regulations still to be published upon which much of the operation of the proposed legislation will depend will simply reinforce the local emphasis of the Bill. From a strategic planning point of view there should be provision made within the Bill for all minerals to be viewed and considered in a regional context and this based upon a deliberate government policy statement pertaining to minerals and land use. Should such government statements ever emerge this would in no sense invalidate the contribution that the Minerals Bill could make, when enacted, to the control and management of minerals operations at a local level.

There are some sections of the Minerals Bill which could be interpreted as offering possibilities for the regional planning, for example that part which pertains to Review Procedures. Although there is no certainty that it was drafted for this purpose, this section of the Bill could be used to facilitate the work of existing and future regional minerals working parties. The Department of the Environment has for a number of years been encouraging county authorities to meet together to consider a range of issues relating to the supply of aggregate minerals and indeed various working papers have been produced (for example, North West Aggregates Working Party, 1979). These reports represent a view being taken of aggregate minerals on a wider regional basis and it would not seem to require a substantial leap forward to have the same or similar groups meeting together to consider other minerals on a similar basis.

Working parties such as those established to look at aggregates or as in the case of the North West Region petrochemicals have drawn together representatives from industry, local government and central government departments and thus established the basis for a dialogue to take place between the main participants involved in mineral and hydrocarbon related matters. It would seem that these working parties could perform two main functions in the future. Firstly, they could provide central government (and in the case of aggregate minerals the National Co-ordinating Group, which is already in existence) with an important range of data and opinion about minerals operations in the regions. Secondly, these groups could act as agents of central government, interpreting national policy, formulating regional strategies, providing guidance and in some cases licencing for the future development of minerals within particular regions. This would not prevent the individual local county planning authorities from discharging their statutory development control functions.

Clearly the establishment of any such regional body to exercise control over minerals operations would remove some responsibility from the counties but it should be remembered that if the regional working parties were given additional powers the individual counties would themselves be participants just as they are at the present time. As such they would be involved in the interpretation of central government policy at the regional level. As well as retaining their development control function at the county level they would be involved in the interpretation of central government policy on minerals at the regional level. This move was supported in the evidence submitted by the Brick Development Association to the Stevens Committee. In this way a more coherent, comprehensive and sophisticated planning and development control system for minerals could be developed.

Not all minerals fall into Blunden's 'ubiquitous' category as do aggregates and therefore in the determination of policy not all regions would be involved with all minerals. However, even where minerals are 'localised' there seems to be no reason why their development should not be conditioned by national policies and regional strategies. After all in advocating the adoption of a more co-ordinated approach to minerals planning other land users and uses have to be fully appreciated and considered. Therefore it would seem unwise not to include the 'localised' category of minerals within any future regional strategies for mineral development.

Although it is encouraging to witness regional aggregate working parties making progress it is not yet clear to what extent issues pertaining to the future development of aggregate minerals will be woven into broader land use planning strategies at the regional level. Clearly those responsible for the formulation of regional aggregate working party papers are aware of the conflicts which are likely to be generated by proposals for the development of minerals; this is also a matter of concern for structure planning authorities. Although there is a presumption against the use of better grade agricultural land and areas of high landscape value at public inquiries dealing with applications to develop minerals considerable pressure is exerted on the one hand to release land for such development and on the other refuse planning permission. As discussed earlier the terms of the planning inquiry make it virtually impossible for broader issues to be considered a problem which would in part be overcome by the adoption of a planning inquiry commission, but it would seem that the only logical solution to this problem is to have a broader and more integrated land use planning context in which local planning decisions can be set. Industry needs minerals, the nation needs prime agricultural land, people in an urban based society need areas for recreation and relaxation: commonsense would insist that wise decisions can only be made in the light of an integrated set of policies for land use which are related to economic, environmental and social factors.

The report of the Stevens Committee (1976) appreciated the need for wider consultation in areas where there was likely to be substantial quarrying and mining because of fears of sterilisation of mineral resources. They suggested the creation of 'Mineral Consultation Areas' a concept accepted by government in its response to their report. Zuckerman (1972) talked of 'Protected Areas' where minerals developments would not normally be allowed and a group of the Sandford Committee (1974) suggested that there should be 'National Heritage Areas' established in National Parks to be kept free from substantial and inappropriate new development. These suggestions clearly reflect the degree of anxiety felt by different interest groups. Simply to have such areas designated would not necessarily bring any greater degree of logic to the process of land use planning and perhaps be no more than a pragmatic response to a confused situation. However, if such areas were designated as part of an integrated land use planning strategy then their contribution could be considerable.

An opportunity to achieve a far greater degree of efficiency and integration in land use planning seems to have been lost during the past twenty years. The guidance that regional strategies should have provided for the preparation of structure plans and the lead that these in turn should have given in the preparation of local and subject plans does not always seem to have been carried through and in many cases this would seem to apply to minerals. It would seem therefore that efforts should be made to try and rescue this situation in order to reduce political pragmatism in land use planning and to instigate the development of regional land use policies which can guide and control future development including mineral extraction.

ALTERNATIVE PATTERNS OF OWNERSHIP AND CONTROL

The suggestion that all minerals should be nationalised would pose all but the most committed socialist government with ideological, management and organisational problems. There would be opposition from many quarters but particularly from those engaged upon working minerals and from those who held mineral rights and who hoped to benefit from them financially. Apart from such opposition, any government would inevitably be drawn into the contentious area of compensation payments and this could prove to be costly.

It is important when considering the possibility of nationalising all mineral resources not to draw too many inferences from the energy minerals which have been nationalised. This process was not embarked upon solely as an act of ideology. For example, when nationalisation took place in the coal mining industry it was also to provide for better organisation and investment. The industry which had become fragmented and highly under capitalised was on the verge of being unable to supply vital energy. If it had not taken place coal mining as a key industry would have declined dramatically exposing the nation to economic and military vulnerability. The social consequences of a further decline in coal mining might have been worse than in the inter war period.

The control exerted by central government over on and offshore hydrocarbon resources indicates their overwhelming importance to the economic well-being of the nation. It is likely that any political party in power would have taken all necessary steps to control the rights over oil and natural gas. Although there have been fierce debates in Parliament as to how government should exercise control over hydrocarbons, the question of these mineral rights being vested in the Crown has never been the subject of significant political acrimony. A distinction should be made between government owning the mineral rights for all hydrocarbons and government through its agencies, actually developing the resources themselves. It is over the issue of how the reserves should be developed that controversy has arisen. Whilst not relinquishing its mineral rights, the Conservative Government which came to power in 1979 has caused some reduction in the operations of the British National Oil Corporation in the North Sea and forced British Gas to sell its profitable holding in the Dorset Wytch Farm gas field. It is not the intention here to develop a political argument about the longer term implications of public versus private control over hydrocarbons other than to suggest that there is little point in government agencies bearing the brunt of development costs only to be forced to sell their holdings when operations become profitable.

The benefits that would accrue from the nationalisation of all privately owned mineral rights would clearly not be worth the costs in either political or fiscal terms. It is difficult to imagine a future situation where the minerals industry could become so fragmented that government was forced to intervene for reasons of national security, social or economic well being. Control over minerals could be more effectively secured by a national licencing system such as that which exists for hydrocarbons. If the government were to operate such a national licencing system for all minerals it might be possible to use it as a mechanism for preventing the unnecessary development of some minerals, a point raised initially in the Scott Report (1942), and as a means of ensuring that national and regional needs for various minerals were both adequately and sensibly satisfied.

Although mineral operators might be horrified at the prospect of a licencing system it has been used to reasonable effect in other sectors. For example, in the agricultural industry Marketing Boards have effectively controlled production for certain goods by using a quota or licencing system (Coppock, 1971). Clearly the opportunities for rationalising production in agriculture might be more speedily and painlessly introduced than in other industries where the capital investment is often higher and the possibility for changing 'crop' are virtually non existent.

If a licencing system were to be introduced for mineral production the issue of permissions already granted would have to be given careful consideration. Here the Minerals Bill might be of some help because the provisions of this proposed legislation allow for all mineral workings to be reviewed and planning permission to be revoked or modified. Although the Bill suggests a new approach to the assessment of compensation clearly payments would have to be made in most circumstances. However the cost of compensation might be acceptable in the context of achieving the goals set out in any future regional strategy for mineral development. The licencing system would be a means of exercising strategic control over mineral extraction to the end of ensuring that the maximum economic benefit might be derived from the country's mineral resources over planned periods of time with the least environmental and social disruption.

CO-ORDINATION AT INTERNATIONAL LEVEL

Any suggestion that the government should relinquish sovereignty over its mineral resources would be viewed with dismay in many quarters. However, since becoming a member of the European Economic Community, Britain has become subject to greater levels of international control and co-ordination in many areas of practice, production and trade. As far as the energy minerals are concerned there have been discussions about the development of common policies. It would seem appropriate to consider the benefits that might accrue from international co-ordination as a logical extension of the ideas which have already been advanced for encouraging inter regional co-ordination within individual countries as a means of ensuring that the best use is made of their mineral reserves.

If the benefits of developing a common minerals policy could be demonstrated to the European Parliament and to the Council of Ministers then the machinery for dealing with a common policy would not be difficult to set in motion within the European Commission. Certainly there is ample experience of formulating and operating common policies within the Community.

Under the provisions of the Common Agricultural Policy little effort has been made to rationalise production, in that the vast majority of the budget is directed towards guaranteeing prices and only a small proportion to agrarian reform. One of the main stumbling blocks has been the reluctance of politicians and ministers from some of the member states to commit themselves to any programme of rationalising food production in Europe. The point of introducing this agricultural example is to highlight how, without the political will to introduce reforms and rationalisation, there is vast over production and a waste of resources. However, the situation in other sectors is not altogether gloomy. For example, strategies have been formulated to try and rationalise production in the steel industry which is a major consumer of mineral resources. Proposals were put forward by Viscount Davignon in 1978 for a longer term plan for the Community's steel industry. As the crisis in the steel industry has heightened, the European Commission with the approval of the Council of Ministers, is likely to intervene soon using its powers under Article 58 of the 1951 Paris Treaty and enforce production quotas. It is of interest to note that the same attitude could be adopted for coal under the provisions of Article 58 of the Paris Treaty. Therefore it can be possible to use common policies and agreements to control production.

How appropriate would be greater European control over the minerals industry is open to some debate. Certainly from an environmental point of view guidance and direction from the Community could be of benefit in land reclamation work or in reducing social hardship and deprivation consequent upon the decline of quarrying and mining in some of the older industrial areas. The effectiveness or worth of Community control over the volume of production is

however open to speculation.   Given the high costs of transporting many
minerals there has been a tendency to use them locally wherever possible.
However, not all minerals have a low unit value and may be economically
transported over greater distances.   This has already been illustrated in the
case of coal.   Hay (1976) has demonstrated 'unit value' is not an independent
variable and is affected by factors of supply and demand.   If mineral
resources which were once widely available become scarce, either through
absolute shortages or as a result of restrictive policies, it may become
economically feasible and perhaps strategically important for many more minerals
to trade on an international market.

If Britain is to remain a member of the European Economic Community,
which hopefully will evolve as a more cohesive unit, then the opportunities
for the international transfer of mineral products increase.   The logic of
advocating a comprehensive approach to minerals planning for regions within
member states must surely be applicable on a wider European scale.   The object
would still be to use minerals wisely and for the good of society as a whole and
it would be none the less important to take account of the full range of economic,
environmental and social implications of any development.   Mineral
development could also figure as an important part of the programmes submitted
by all member states under the European Regional Development Policy.
Furthermore, European funding both from the European Coal and Steel Community
and the European Investment Bank makes an important contribution to the
development of coal reserves in the United Kingdom and there may well be
similar possibilities for helping other sectors of the economy in which minerals
play a role.

12 Prospects and Prescriptions

Minerals planning in the United Kingdom has evolved from what was, in the 1940's, an often imprecise and unsure attempt to control the use of land for the winning of minerals, into a mature and highly professional activity. The increasing level of geological information, the growing awareness of operators that environmental considerations must be taken into account and the growing sophistication of planning methodology, have helped to resolve many of the problems that typified the immediate post war era. Chapter ten of this text indicated the nature and incidence of the major problems that remain and which represent the priorities towards which any action should be directed. Whilst it is relatively easy to create an alternative brave new world for minerals planning, the present authors are aware that drastic or radical innovations are unlikely to be acceptable to either central government or industry. In Chapter eleven some alternative approaches and policies were put forward. Some of these may, in the near future, become realities due to the actions of nations, organisations and individual companies outside the competence of any British government. In the field of energy, events external to the United Kingdom have done much to influence the way in which national and local planning have considered new and often unfamiliar constraints upon their freedom of action. From this example it is clear that strategic planning has to adapt in order to ensure that a full range of future options can be taken into account.

The somewhat chequered history of strategic planning in the United Kingdom can only provide limited comfort to those who are faced with new realities. This uneasy situation is all the more serious in the case of minerals planning. Despite the excellent preparatory work done by the Stevens and Verney Committees, little has yet emerged from Whitehall to suggest that a positive and dynamic role for minerals planning will be easily achieved. Stress has to be laid upon the positive and developmental functions that lie within the capability of the planning system, but to mobilise such potential requires explicit strategic guidance from central government and co-operation and a clear statement of intentions from the minerals industry. Central government has been unwilling to provide the 'special regime' called for by the Stevens Committee. So be it. In order to move forward into an age when shortages of key minerals will increasingly manifest themselves will require all the ingenuity, adaptability and resourcefulness that can be gleaned from the past thirty five years experience of attempts to plan for minerals.

This final Chapter attempts to provide guidance, based on the lessons that have now been learned, for the future of strategic minerals planning. It is a modest and partial prescription and represents the first steps in what will hopefully become a full review of the problems that face the activity of minerals planning in forthcoming years. The authors make no apology for considering their suggested reforms in a somewhat hierarchical sense, for that is the reality of government and the planning system. A number of additional points are also made which relate more generally to the ways in which strategic planning for minerals might evolve during the next twenty years.

NATIONAL PRIORITIES

It is apparent that when operators prepare their proposals and when applications are considered by a planning authority or at an inquiry, there is little by way of policy guidance to which reference can be made. As has been shown, this policy vacuum generates delay in the consideration of applications and can cause decisions to be made which do not always represent the best possible outcome for society as a whole. National policies for individual minerals are urgently required, especially for those resources where there is either a possibility of early exhaustion or where major new expansions of production are likely. The need for such policies was clearly supported in much of the evidence submitted to the Stevens Committee. British Industrial Sand expressed this need very clearly, 'as resources in areas of low amenity are exhausted operators will be forced to seek planning permissions in areas of higher amenity and environmental quality'.)Stevens,1976). It is vital that such policies are related to other matters, especially national policies for the use of land, the social and economic objectives of government, major environmental priorities and the future development of the transport system. The question of efficient transport is a crucial matter for minerals planning in that it is likely that significant inter regional transfers will persist. The case for the generation of policies for individual minerals is that they will ensure a continuous supply of materials to industry whilst minimising the social and environmental costs which will inevitably be incurred. As well as ensuring the provision of necessary minerals, policy should also have regard to the potential role of mineral extraction and processing in the economic growth of the nation. This may mean in a limited number of cases, delaying a project in order to allow for a full assessment of economic potential and to enable the full participation of British industry in the project.

The development of national minerals policies will ensure that, in general terms, the conflicts which occur between and within the various branches of central government can be resolved. In addition, the residents of an area which becomes subject to mining claimed to be in the national interest, will be able to be reassured that such a claim can be substantiated. Furthermore, a set of explicit policies will expedite the work of strategic planning authorities and could reduce the delays which are common at public inquiries. In the case of certain key minerals this need is urgent; an upturn in demand for aggregates from the construction industry could result in acute shortages of supply whilst energy minerals of all kinds require a clear indication of the long term intentions of government. The question of the lead time required for the development of specific minerals resources necessitates the adoption of different planning horizons.   However, as a first step it should be possible to plan as far ahead as fifteen years.   This timescale is already used by many operators and by regional aggregates working parties.

From these national policies it should be possible to identify areas and sectors where conflict or low productive performance is likely to emerge.   This may require choices to be made on political, social or military grounds but at least the process of choice will be informed and the outcome of the exercise can be anticipated.   What should be avoided is the preparation of a national plan for minerals.   Such blueprint plans have the unfortunate habit of becoming rapidly out of date and constraining the process of adaption and innovation.   In case this may seem to be an argument for total flexibility and a recipe for constant confusing and damaging changes of direction, it should be stated that what is being advocated is the introduction of robust policies.   These policies should be sturdily built, anticipatory in design and able to withstand future shocks without their wholesale abandonment.   A set of policies produced on this basis would provide the minerals industry with the degree of certainty that it needs in order to progress its corporate planning and investment programming and would engender a clear sense of direction in the activities of planning authorities.

Other benefits would be derived; the inquiry system might be relieved of some of the problems which at present beset it and our knowledge of the occurrence and extent of mineral resources might be increased.   The latter point is considered here, the former will be dealt with later.   As mining activities move into new areas, much detail is added to the geological map of Britain. Some of this information is obtained by private exploration and is subject to the constraint of commercial confidentiality.   If government's views on the extraction of a particular mineral were clear, then the present unwillingness of many operators to disclose survey information might be diminished.   It should be the intention of central government to assemble the fullest possible picture of the nation's mineral wealth and as well as utilising private survey results it should also encourage and finance an extension of the work of the Institute of Geological Sciences.   A further benefit from the development of policies and an extension of geological knowledge would be the development of licencing arrangements for minerals which lie within the control of central government. Whilst it is apparent that policy for the hydrocarbons industry has benefited from the lodging of well records with the Department of Energy and from the development of explicit policies for the extraction of oil and gas, it is also clear that the creation of a licencing system has provided potential benefits in ensuring that extraction can occur more easily in those areas which government wishes to develop quickly and encouraging exploration at a lower licence cost elsewhere.

Least the argument should become too theoretical it is worth pointing to the advances made in recent years by two arms of government in providing policy guidance. The Scottish Development Department has attempted, through a series of national guidelines, to create a policy context for minerals planning in Scotland. To an extent these efforts have been inhibited by a lack of progress in Whitehall though they do show what can be achieved given foresight and a desire to set down explicit strategic priorities. Elsewhere, the early years of the Department of Energy were most productive in clarifying the aims of national energy policy, although in this case insufficient attention was paid to many of the social and environmental aspects of the exploitation of coal and hydrocarbons.

In another sense government needs to consider how it wishes policy to be translated into action. Here it should provide a more sophisticated and detailed level of technical advice to the authorities responsible for lower level planning functions. The spatial incidence of many minerals means that the same deposit may occur on both sides of an administrative boundary. If national policies are to instill confidence in operators, then they will have to be interpreted by all authorities in the same way. This is not a proposition for bland uniformity, rather an argument for a consistent and co-ordinated approach to be adopted by strategic planning authorities. Matters which require the attention of central government include the preparation of a fuller national standard application form; this would follow from current attempts at devising standard regional forms, the specification of assessment procedures, the revision of the present Green Book to take account of strategic matters and the requirement that all authorities prepare a minerals plan. There has been considerable debate on the question of introducing a standard national mineral application form. The questions which are asked should be broadly those suggested in Annex 6A of the Stevens Report (1976) with the addition of detailed questions on the economic, environmental and social consequences of a proposed working and further information on the secondary impacts of the mining operation. Such a form would expedite the processing of a planning application, especially if it was completed following detailed discussions between the applicant and the relevant planning authorities. It would also provide the basis for, and in turn would be responsive to, the modified assessment procedure which is suggested below.

If the proposed European directive on environmental impact assessment were not to be accepted or agreed, there is still a need to modify the present methods of evaluation and assessment which are utilised by planning authorities. The role which more comprehensive forms of assessment can play in producing speedier and better information on which decisions can be based has been discussed earlier. It has been argued by other authors (Murphy, 1979) that, in the case of minerals planning, the adoption of the Dobry recommendation for the more detailed assessment of major or controversial applications under a Class B procedure, allied to the use of a standard application form as suggested by Stevens, would enable a fuller assessment to be undertaken of the majority of proposals for the extraction of minerals. What is needed therefore is the provision of guidance to local planning authorities on when and how to operate a system of comprehensive and co-ordinated assessment. The approach suggested by the Aberdeen University team (Clark, Chapman, Bisset and Wathern, 1976) has much to commend it, although certain modifications would be necessary to ensure that the approach is appropriate to the assessment of proposals for the extraction of minerals and the estimation of the effects of multiplier and secondary development.

The above matter could be satisfied by a series of amendments to the Green Book. The current revision of this document is not expected by the present authors to produce a set of guidelines that would effectively require authorities to institute a system of environmental impact assessment. A more serious issue is that current amendments are unlikely to result in the stimulation of strategic planning activity, indeed a reinforced development control system would result in a greater degree of inertia and would reinforce the essentially negative aspects of local planning activities. The provision, as by the Scottish Development Department, of strategic planning guidelines would shift the emphasis towards forward looking and positive planning.

A further impetus could be given to this positive approach if all strategic local planning authorities were required, by the Department of the Environment, to prepare minerals subject or local plans. Where such plans have emerged they demonstrate what can be achieved and represent a willingness to fully consider the range of issues that are associated with mineral resources. They are an attempt to escape from harmful incrementalism whereby each application is considered on its own merits. With an increasing awareness of the crucial role that minerals have to play in the economic future of the United Kingdom, and recognising the wider social and environmental consequences of mining, then positive forward planning becomes a necessity. The incremental approach may be applicable to small sand or gravel pits (although the authors would cast doubt even upon this), but is clearly inappropriate to proposals for the extraction of major coal, hydrocarbon, stone, or metaliferous mineral resources. In addition, such plans provide an improved data base and help to avoid the unecessary sterilisation of reserves.

These national priorities for action are modest and they represent the basis that has to be laid by central government if it wishes to encourage the responsible development of minerals. As can be seen from these proposals, there are no unrealistic suggestions for major additional legislation or the creation of a vast new bureaucracy. The following discussion of lower tier reforms assumes that the national priorities will be attained.

REGIONAL FUNCTIONS

Although the Stevens Committee came very close to recommending the creation of a regional system of minerals planning, in their final conclusions they gave reserved support to the retention of minerals as a county function. However, the arguments that were advanced by those who gave evidence to Stevens, for the creation of regional minerals authorities were powerful. The range of witnesses who supported the regional approach included the minerals industry (for example, the Brick Development Association), organisations concerned with environmental conservation (The Committee for Environmental Conservation) and consultants (Mackay and Schnellmann Ltd.). A regional approach was also supported by much of the work of the Verney Committee, who saw the pressures of supply and demand for aggregates as a regional phenomenon.

Whilst it is undoubtedly true that many county authorities perform their minerals planning function excellently, other counties lack the necessary specialist staff. The operation of the regional aggregates working parties provides evidence that the whole is greater than the sum of its constituent parts, that is, that their understanding of the problem of regional patterns of transport and supply and demand is greater than that of any individual county authority. Furthermore, certain of the working parties have expressed the view that aggregates should be considered on the basis of service areas, many of which cross county boundaries. Once more, the operation of regional systems in Scotland adds weight to the argument for a regional strategic approach. In the case of other minerals a regional approach might aid in the consideration of alternative sites, anticipation of the transport requirements of operators and in the assessment of the ancilliary impacts of mineral working whi ch are often spread across a number of local authority areas. It is clear, for example, that the direct effects of oil developments in Scotland were not confined to the Highlands, Grampian and Tayside Regions. In North West England many opportunities for the maximisation of the developmental effects of Morecambe Bay gas were lost due to the lack of an agreed regional strategy. With the extension of coal working into the southern part of the West Midlands a number of county authorities will be involved. In order to provide details which will assist in the preparation of national policies for all minerals, and to ensure that specialist staff are available to expedite the processing of applications and the monitoring of changing circumstances, regional authorities are essential. The secondment of county officers to regional minerals teams need not result in extra expense, but it would provide a better service for industry and may well hasten the generation of strategic minerals plans.

As was stated in the Department of Industry's response to the Stevens Committee, minerals play a vital role in the economy. The minerals industry also provides a significant amount of direct and indirect employment. On this basis, and recognising the often widespread social and environmental effects of mining, regional minerals teams should work within the context of a continuing system of regional strategic planning. Although many of the regional strategies of the 1960's and 1970's gave, in the opinion of the authors, insufficient attention to minerals issues they did provide an essential basis for the assessment of integrated minerals development. The beneficial effects of taking an integrated view have already been noted. Essentially such a view would engender a positive approach which would recognise the importance of minerals to the economy of regions, rather than regarding them as troublesome land use problems where each case has to be considered on its own merits.

LOCAL AUTHORITIES AND THE MINERALS INDUSTRY

At present county planning authorities have responsibility for the assessment of applications for the extraction of minerals and the preparation of statutory strategic plans. As can be seen from the preceeding observations this allocation of functions to counties is not a cause for complacency nor, in some cases, optimism for the effective future development of the minerals industry. Much has been written criticising the slow operation of planning assessment. Equally, county authorities have accused operators of failing to disclose important information. The acrimony which surrounds many minerals applications has also pervaded public inquiries. In short, the situation is one where increasingly large and sophisticated operators, drawing upon the expert advice provided by consultants, require county authorities to provide a speedier assessment of their proposals. This is occurring at a time when local authorities face staff shortages and have to increasingly consider the environmental and social objections voiced by their residents.

If county authorities are to continue to discharge their minerals planning function it is important that they are provided with adequate advice and guidance by central government. Questions concerning national policies, standardised application forms, assessment procedures and planning guidelines have already been discussed in terms of the functions of central government. It is important to relate such matters to the functioning of local authorities.

Much of the heated argument that is invoked by applications which are claimed to be in the national interest, would be dispelled if detailed national policy statements were available. Industry wishes to know the limits that constrain its freedom of action, otherwise much time and expense can be wasted. Counties may think that their view is a fair representation of the national interest but it may be based on the most hazy of statements from central government. In the preparation of both structure and subject plans a national statement of objectives is a prerequisite. Differences of interpretation will remain, but at least the basic facts will have been established.

Assessment procedures differ between local authorities. This can be explained on grounds of staffing, the geographic endowment of areas and the social and economic composition of the population. What is clear is that operators face different information requirements from authorities and that their proposals are processed in a number of ways. There are of course circumstances in which special procedures will be required. However, in order to hasten the process of assessment and to ensure that a full and fair account is taken of all major consequences, standardised assessment procedures should be utilised by all authorities. The aim should be to ensure that best current practice becomes the norm. The Royal Town Planning Institute (1980) has claimed that environmental impact assessment has been found, by both local authorities and commercial operators, to accelerate the process of decision making. This text has emphasised the important role of minerals in economic development, equally it is important to stress that good decisions can aid overall economic efficiency. Such efficiency, as was shown by the Hunt Report (1969), can be measured in environmental and social terms. If operators require decisions to be made as quickly as possible then comprehensive assessment can be beneficial; it can also help to avoid the harmful social and environmental side effects of mineral exploitation. A revised version of the Green Book should take account of the potential offered by environmental impact assessment and also provide a model for a minerals application form in order that the information requirements of authorities can be made explicit to operators.

Other matters of national guidance that can assist local authorities and operators include a national specification for the designation of mineral consultation areas and the delimitation of potential areas. In order to avoid the sterilisation of important mineral deposits, consultation areas are a necessity, and the designation of such areas is strongly supported by industry. Expense and time can be saved by operators in the preparation of their plans and local populations would be reassured if areas of high amenity value were to be declared as protected areas. The designation of both types of area can only proceed at a local level if geological and other data are available and if they can be set within the context of detailed and explicit strategic plans.

Although the majority of structure plans include some reference to mineral issues, often the level of detail which is provided is insufficient to guide operators and the development control system. Certain local authorities have prepared subject or local minerals plans. It has been argued previously that the Secretary of State for the Environment should require all authorities to prepare such a plan. The greater knowledge of mineral resources that they would provide is important, but they would also assist in allowing for a more rapid and a better informed assessment of applications. Industry would know where and under which circumstances it could expect to operate and local populations would be provided with a degree of reassurance. There will be difficulties. Blight could ensue if hasty or ill considered proposals were made, but important potential benefits exist.   The creation of a strategic framework of this nature would allow both operators and authorities to anticipate changes in the patterns of supply, the long lead times for many mining developments could be incorporated within a planned approach and the relationship between mining and other land uses would be established.   Long term planning, following the example set by Cornwall County Council (1974 and 1979) and Grampian Region (1980), would clarify many issues which are at present the subject of doubt and uncertainty. The key to effective structure and subject plans is the willingness of industry and county authorities to co-operate and exchange information on their respective views and intentions.

Given the modifications which have been proposed it is likely that a reduced number of applications will have to be determined at inquiry.   It is anticipated that the cases that are in future referred to an inquiry will be those which are the subject of major controversy or where inadequate policy guidance exists. Attention has been given in this text to the failings of the inquiry system; these are mainly that it prolongs the period which is required to produce a decision and also that it is limited in scope by the strictures of legislation.   The alternative, a planning inquiry commission has also been discussed.   Given greater attention to the production of explicit policies for all minerals, then the length of inquiries could be reduced.   Controversial or multiple site applications might be better considered by a planning inquiry commission.   It would expedite matters if all concerned in future inquiries were required to submit written statements which might be tested against national policy.   This could, in some cases, involve providing certain of the parties who are involved with financial or technical assistance.   In the light of the adjustments to the system of minerals planning which have been proposed in this Chapter, inquiries should be able to focus on the key issues of concern.   It is not intended to elaborate further on the future of inquiries.   They represent the irretrievable breakdown of co-operation between the various parties involved in the current system of minerals planning.   Rather it is hoped that modifications to that system will foster a positive and beneficial approach to future planning.

TOWARDS THE FUTURE

The extraction of minerals has had a profound and lasting effect on the landscape of Britain.   Settlement patterns, long obsolete, bear testimony to previous mineral working, whilst abandoned tin mines, steel works and collieries record the passing of previous economic activities.   It is clear that, almost unlike any other major sector with which planning is engaged, minerals have a great future potential for expansion and the direct creation of wealth. On the one hand this potential is constantly being extended through new discoveries and exploration, whilst on the other the stock of certain minerals is eroded by extraction.   Minerals can only be mined once, even given the possibiliti es of reworking old waste tips, the alternatives to the extraction of new deposits are limited. Mining and quarrying are also unique in the sense that they are the only form of development which is destructive by intent, when an ore is extracted then there is no second chance.

Many regional economies have their origins in the chance discovery of a mineral resource, others owe much to the economic rationale provided by basic materials. The transformation to the land brought about by mineral working, once regarded as ugly and despoiled is now regarded as historically interesting, if not scenically attractive. Hoskins observed, 'there is a point when industrial ugliness becomes sublime'. There are also vast scars on the surface of Britain and large areas of despoliation that await reclamation.

The prime objective of minerals planning must be to maximise the economic benefits that can be obtained from the extraction of resources whilst minimising the social and environmental costs which are incurred. No amount of compensation or restoration can make amends for bad planning and decision making. The conflicts of interest which are demonstrated at many public inquiries are founded in real fears and suspicions. The minerals industry is concerned with cost effectiveness and obtaining a clear and rapid response from planning authorities. Such authorities are mindful of their responsibilities to residents and the views of central government. Meanwhile central government maintains an almost inscrutable silence on matters of long term policy and strategic planning.

In the absence of any positive move towards the production of national policies for all minerals then uncertainty will remain. Guidance to local or regional authorities on the best available methods of assessment and plan preparation is urgently required. The decisions which are produced by future minerals planning will never be the subject of universal agreement. Someone will always lose. Central government must take the initiative and move forward towards a more positive and innovative role for minerals planning.

Bibliography

Amey Roadstone Corporation Ltd., Land Reclamation at Sutton Courtney, Amey Roadstone Corporation, Ltd., Oxford 1980.

Ardill, J., The New Citizens Guide to Town and Country Planning, Charles Knight, London, 1974.

Arnold, P. and Cole, I. The Development of the Selby Coalfield, Department of Social Administration and Social Work, The University, York., 1981.

Bain, J.S., Environmental Decay : Economic Causes and Remedies, Little Brown, Boston, Mas., 1973.

Bate, R., 'Topping Up', Minerals Planning, No. 2., 1980,a.

Bate, R., 'The Cost of Planning', Minerals Planning, No. 4., 1980,b.

Berry, D and Steiker, G. 'The Concept of Justice in Regional Planning', Journal of American Institute of Planners, Vol.40, No. 6., 1974

Best, R.H. and Champion, A.G. 'Regional Conversions of Agricultural Land to Urban Use in England and Wales 1945 to 1967'. Transactions of the Institute of British Geographers, No. 49., 1970.

Best, R.H. and Swinnerton, G.S. The Quality and Type of Agricultural Land, Converted to Urban Use, Social Science Research Council, London., 1974

Bigham, D.A., The Law and Administration Relating to Protection of the Environment, Oyez Publications, London., 1973

Blunden, J. The Mineral Resources of Britain, Hutchinson, London., 1975.

Blunden, J. Resource Exploitation, Open University Press, Bletchley, Bucks. 1977.

Blowers, A., The Limits of Power, Pergamon Press, Oxford, 1980.

Bracegirdle, B., The Archaeology of the Industrial Revolution, Heinemann, London., 1973.

Bracey, H.E., People and the Countryside, Routledge and Kegan Paul, London, 1970.

Briggs, R.T., 'The National Water Sports Centre, Holme, Pierre point, Nottingham'., Journal of the Institute of Municipal Engineers, Vol.98., 1971.

Caisley, R., 'Minerals Proposals Give us the Tools but not the Men', The Planner, Vol. 64., No. 6. 1978.

Catlow, J. and Thirwall, C.G., Environmental Impact Analysis, Department of the Environment Research Report. No. II. HMSO., London. 1976.

Central Statistical Office,. National Income and Expenditure 1980, HMSO London., 1980,a.

Central Statistical Office,. 'Gross Domestic Product by Industry Groups'., Economic Trends., No. 325. 1980,b.

Chadwick, Sir Edwin. (Chairman) The Sanitary Condition of the Labouring Population, Poor Law Commissioners, London. 1842.

Cheshire County Council., Salt: A Policy for the Control of Salt Extraction in Cheshire., Cheshire County Council, Chester. 1969.

Ciriacy-Wantrup, S.V. Resource Conservation, Economics and Policies, University of California Press, Berkley, California. 1952.

Civic Trust., Derelict Land, Civic Trust, London. 1964.

Civic Trust., Reclaiming Derelict Land, Civic Trust, London. 1970.

Clapham, Sir John. A Concise Economic History of Britain from Earliest Times to 1750., Cambridge University Press, Cambridge., 1963.

Clark, B.D. Chapman, D.Bisset, R and Wathern, P. Assessment of Major Industrial Applications : A Manual, Department of the Environment Research Report No. 13, HMSO. London., 1976.

Clifford, S. Impact Analysis: A Critical Review of Experience in Britain, Social Science Research Council, London. 1975.

Clifford, S. 'EIA – Some Unanswered Questions', Built Environment, Vol. 4. No. 2. 1978.

Coleman, A. 'Land Reclamation at a Kentish Colliery', Transactions of the Institute of British Geographers, No. 21. 1955.

Commission of the European Communities, Draft Directive Concerning The Assessment of The Environmental Effects of Certain Public and Private Projects, COM(80) 313 Final, Commission of the European Communities, Brussels., 1980.

Commoner, B. The Closing Circle: Confronting the Environmental Crisis, Jonathan Cape, London. 1971.

Coppock, J.T. An Agricultural Geography of Great Britain, Bell and Sons, London. 1971.

Cornwall County Council, The St. Austell China Clay Area, Cornwall County Council, Truro. 1974.

Cornwall County Council, The St. Austell China Clay Area Short Term Development Plan 1977-1984. Cornwall County Council, Truro., 1979.

Cornwall County Council, County Structure Plan, Cornwall County Council, Truro., 1980.

Council for Environmental Conservation, Scar on the Landscape, Council for Environmental Conservation, London. 1979.

Cottrell, A. Environmental Economics, Edward Arnold, London., 1978.

Darby, H.C. 'The Clearing of the Woodland in Europe' in Thomas, W.L.(ed) Man's Role in Changing the Face of the Earth, University of Chicago Press, Chicago., 1956.

Dasmann, R.F. Environmental Conservation, John Wiley and Sons, New York., 1976.

Davies, T. The Historical Context of Regional Planning in Caernarvonshire, School of the Environment, Polytechnic of Central London, London., 1974.

Dean, F.E. To See or Not to See – Some Environmental Aspects of the Onshore Natural Gas Programme, The Institution of Gas Engineers, London. 1974.

Dent, D. 'Problems of Farming on the Urban Fringe', Urban Fringe Farming Seminar, Ministry of Agriculture Fisheries and Food, Newcastle upon Tyne, 1979.

Department of Employment., 'Annual Census of Employment; June 1976, Department of Employment Gazette, December, 1977.

Department of Energy, Coal for the Future: Progress with Plan for Coal and Prospects for the Year 2000, HMSO, London. 1977.

Department of Energy, Development of the Oil and Gas Reserves of the U.K. HMSO, London. 1980.

Department of the Environment, Development Involving Agricultural land, Department of the Environment, London. 1976,a.

Department of the Environment, Forecasts of the Demand for Aggregates., Department of the Environment, London. 1978,a.

Department of the Environment., Regional Aggregate Working Parties : Collation of Interim Reports, Department of the Environment, London. 1978,b.

Department of the Environment,. Peak District National Park Structure Plan : Proposed Modifications by the Secretary of State, Department of the Environment, London. 1978,c.

Department of the Environment., Response by the Secretaries of State to the Report of the Verney Committee. Department of the Environment,London. 1978,d.

Department of the Environment, . Statement of the Conclusion of the Secretary of State for the Environment on the Report of the Committee on Planning Control over Mineral Working, Department of the Environment, London,1978,e.

Department of the Environment, Consultation Paper on Guidelines for Aggregates, Department of the Environment,London. ,1979.

Department of the Environment, Regional Aggregate Working Parties : Collation of the Results of the 1977 Survey, Department of the Environment, London., 1980,a.

Department of the Environment, Development Control Policy and Practice, Department of the Environment, London. 1980,b.

Department of the Environment, Town and Country Planning : Development Control Functions, Department of the Environment, London. 1981.

Department of the Environment and Welsh Office. Local Plans: Public Local Inquiries., HMSO, London. 1977.

Dodd, P. 'Coal versus Crops : Prime Farm Land in Danger', Farmers Weekly, April 6. 1979.

Down, C.G. and Stocks, J. Environmental Impact of Mining, Applied Science Publishers, London. 1977.

Down, C.G. Stocks, J and Pryor, R.N. The Environmental Impact of Large Stone Quarries and Open Pit Non Ferous Metal Mines in Britain, Department of the Environment Research Report, No. 21. HMSO,London. 1976.

Durham County Council. County Development Plan 1951, County Council of Durham, Durham. 1951.

Durham County Council, County Development Plan Amendment 1964, County Council of Durham, Durham. 1964.

East Anglia Regional Strategy Team, Strategic Choice for East Anglia, HMSO, London. 1974.

Eve, J.R. Town and Country Planning (Minerals) Bill. J.R.Eve, London. 1981.

Fletcher, J. 'The Impact of Mineral Operations on Local Residents', Paper presented at seminar on Minerals Planning, Rewley House, Oxford, 1980.

Friends of the Earth, Tunstead Quarry, Evidence in respect of the appeal by I.C.I. against the refusal of planning permission to extract minerals at Old Moor, Department of the Environment, London. 1976.

Grampian Regional Council, Contingency Plan for Petrochemical Industries, Grampian Regional Council, Aberdeen, 1980.

Gregory, R. The Price of Amenity, Macmillan, London. 1971.

Griffiths, M. 'Monumental Delays', Minerals Planning, No. 2. 1980,a.

Griffiths, M. 'New Code of Practice suggested for Minerals Operators', Minerals Planning, No. 5. 1980,b.

Hampshire County Council, Policies for Hydrocarbon Development in Hampshire, Hampshire County Council, Winchester, Hants. 1981.

Harrison, M and Machin, S. Mineral Planning Appeals in Great Britain, Mineral Planning Publications, Northallerton, North Yorkshire, 1981.

Hay, A.M. 'A Simple Location Theory for Mining Activity', Geography, Vol.61, No. 2. 1976.

Henry, J. 'Minerals Planning in Lothian', Minerals Planning, No.2. 1980.

Hills, P. 'Belvoir the Crucial Test Case', Planning, No. 308. 1979.

Hilton, K.J. (ed) The Lower Swansea Valley Project, Longmans, London. 1967.

Hoskins, W.G. The Making of the English Landscape, Hodder and Stoughton, London, 1955.

Humble, J. Social Responsibility Audit, Foundation for Social Responsibility in Business, London. 1973.

Hunt, Sir Joseph, (Chairman), The Intermediate Areas, HMSO, London. 1969.

Hutber, F. 'Macro Regional Impact of Indigenous Energy', Regional Studies Association Conference Paper, Edinburgh, 1980.

Imber, V. Public Expenditure 1978-79, Outturn compared with Plan, Government Economic Service, London.1980

Institute of Geological Sciences, United Kingdom Mineral Statistics, HMSO, London., 1979.

Institute of Geological Sciences, United Kingdom Mineral Statistics, HMSO, London., 1980.

Jackson, M.P. The Price of Coal, Croom Helm, London. 1974.

Jelley, A.D. 'Town and Country Planning (Minerals) Bill' Journal of Planning and Environment Law, March. 1981.

Joint Monitoring Steering Group, A Developing Strategy for the West Midlands, West Midland Regional Study, Birmingham, 1979.

Lake, J.R. Unburnt Colliery Shale - Its Possible Use as Road Fill Material, Road Research Laboratory, Crowthorne, Berks.

Lambert, W.M. Streetley Brick Ltd., Private correspondence. 1981.

Lee, N. and Wood C.M. Methods of Environmental Impact Assessment For Major Projects and Physical Plans, Commission of the European Communities, Brussels

Leivers, R. Evidence Presented by Chief Assistant Land Agent at the Vale of Belvoir Public Inquiry. Leicestershire County Council, Leicester, 1980.

Lewis, T.N. and McNicoll, I.H. North Sea Oil and Scotlands Economic Prospects, Croom Helm, London. 1978.

Ley, G.M.M. Proposed Mineral Extraction at Old Moor, Appeal by I.C.I. Proof of Evidence. Department of the Environment, London. 1976.

Lichfield, N. 'Evaluation Methodology of Urban and Regional Plans : A Review', Regional Studies, Vol, 4. No. 2. 1970.

Lloyd, P.E. and Dicken, P. Location in Space, Harper and Row, London, 1977.

Machin, S. 'Selby Coalfield Rail Diversion', Minerals Planning, No. 2. 1980.

McAuslan, P. The Ideologies of Planning Law, Pergamon Press, Oxford, 1980.

McDonald, S.T. 'The Regional Report in Scotland', Town Planning Review, Vol. 48, No. 3. 1977.

McLoughlin, J.B. What is Strategic Planning? Discussion Paper, Department of Town and Country Planning, Liverpool Polytechnic, Liverpool. 1978.

McVean, D.N. and Lockie, J.D. Ecology and Land Use in Upland Scotland, Edinburgh University Press, Edinburgh. 1969

Manners, G. Keeble, D., Rodgers, B and Warren, K., Regional Development in Britain, John Wiley and Sons, London. 1972.

Mead, A. 'Tungsten in Devon', Minerals Planning, No. 4. 1980.

Merseyside County Council, Merseyside Structure Plan, Merseyside County Council, Liverpool, 1979.

Miles, D. 'The Archaeological Implications of Mineral Extraction', Paper presented at seminar on Minerals Planning, Rewley House, Oxford, 1980.

Ministry of Agriculture, Fisheries and Food , Food From Our Own Resources HMSO, London, 1975.

Ministry of Agriculture, Fisheries and Food, Farming and the Nation, HMSO London, 1979.

Ministry of Housing and Local Government, The Control of Mineral Working, HMSO, London, 1960,a.

Ministry of Housing and Local Government, Review of Development Plans, Ministry of Housing and Local Government, London. 1960,b.

Ministry of Housing and Local Government, The Use of Conditions in Planning Permissions, Ministry of Housing and Local Government, London, 1968.

Ministry of Town and Country Planning, The Control of Mineral Working, HMSO, London, 1951.

Miernyk, W.H. 'Coal and the Future of the Appalachian Economy', Appalachia, Vol.9, No. 2. 1975.

Mishan, E.J. The Costs of Economic Growth, Penguin Books, London, 1967.

Moore, W.E. A Dictionary of Geography, Penguin Books, London, 1963.

Moses, K. 'New Mines in the Countryside', Chartered Surveyor, October,1980.

Murphy, R.F. 'Environmental Impact Analysis: Its Relevance for the U.K's Extractive Industries", Minerals and the Environment, Vol,1.No.1. 1979.

National Coal Board, Compensation for Mining Subsidence, National Coal Board, London, 1957.

Nature Conservancy Council and National Environment Research Council, Nature Conservation in the Marine Environment, Nature Conservancy Council, Shropshire, 1979.

North, J. and Spooner, D.J. 'The Geography of the Coal Industry in the United Kingdom in the 1970s : Changing Directions', GeoJournal, Vol.2. No. 3. 1978.

North West Aggregates Working Party, Second Report, Cheshire County Council, Chester, 1979.

Northern Region Strategy Team, Strategic Plan for the Northern Region, HMSO, London, 1977.

Olschowy, G. 'Case Study : Industry and Landscape', Paper presented at the United Nations Conference on the Human Environment, Stockholm, 1971.

O'Riordan, T. Environmentalism, Pion Press, London. 1976.

Orrell-Jones, K. 'Aggregates for the South East', Quarry Management and Products, April. 1978.

Orwell, G. The Collected Essays, Journals and Letters of George Orwell, Volume I, An Age Like This, Penguin Books, London, 1970.

Oxenham, J.R. Reclaiming Derelict Land, Faber and Faber, London, 1966.

Pearce, D., Edwards, R. and Harris, T., The Social Incidence of Environmental Costs and Benefits, Social Science Research Council, London, 1979.

Peak Park Joint Planning Board, Peak District National Park Structure Plan,. Peak Park Joint Planning Board, Bakewell, Derbyshire, 1976.

Pocock, D.C.D. 'The Novelist's Image of the North', Transactions of the Institute of British Geographers, Vol. 4., No. 1. 1979.

Rees, T.L. 'Population and Industrial Decline in the South Wales Coalfield', Regional Studies, Vol. 12. No. 1. 1978.

Regional Aggregates Working Parties, Collation of Interim Reports, Department of the Environment, London. 1978.

Regional Aggregates Working Parties, Collation of the Results of the 1977 Survey, Department of the Environment, London. 1980.

Rio Tinto Zinc Corporation Ltd., Annual Report and Accounts 1977, Rio Tinto Zinc Corporation Ltd., London. 1977.

Rio Tinto Zinc Corporation Ltd., Modern Mining and the Environment, Rio Tinto Zinc Corporation Ltd., London. 1978.

Road Research Laboratory, Unburnt Colliery Shale - Its Possible Use As Road Fill Material, Technical Note 317, Road Research Laboratory, Crowthorne, Berks., 1968.

Roberts, D.W. An Outline of the Economic History of England, Longmans, London. 1948.

Roberts, M. An Introduction to Town Planning Techniques, Hutchinson, London, 1974.

Roberts, P.W. and Shaw, T. 'The Planning Consequences of North Sea Oil; The Impact upon Scotland', Urbanistica, No. 67. 1977.

Roberts, P.W. and Shaw, T. 'Onshore Planning Implications of the offshore Development of Mineral Resources', Marine Policy Vol.4, No.2. 1980.

Rodmell, G.A. 'A Special Regime for Minerals'. Journal of Planning and Environment Law, June. 1976.

Royal Town Planning Institute, Coal and the Environment, Royal Town Planning Institute, London. 1979.

Royal Town Planning Institute, Environmental Assessment of Projects, Royal Town Planning Institute, London. 1980.

Rowan-Robinson, J. Some Misconceptions About Major Inquiries, Department of Land Economy, The University of Aberdeen. 1980.

Sandbach, F.R. 'The early campaign for a National Park in the Lake District', Transactions of the Institute of British Geographers, Vol.3.No.4. 1978.

Sandford, Lord, (Chairman) Report of the National Park Policies Review Committee, HMSO, London. 1974.

Scott, Lord Justice, (Chairman) Report of the Committee on Land Utilisation in Rural Areas, HMSO, London. 1942.

Scottish Development Department, Coastal Planning Guidelines, Scottish Development Department, Edinburgh, 1974.

Scottish Development Department, National Planning Guidelines : Aggregate Working, Scottish Development Department, Edinburgh, 1977.

Searle, G. 'Copper in Snowdonia National Park', in Smith, P.J. (ed), The Politics of Physical Resources, Penguin Books, London, 1975.

Shane, B.A. Estimating the Demand for Aggregate in Great Britain, Transport and Road Research Laboratory, Crowthorne, Berks. 1978.

Sheail, J. Rural Conservation in Inter War Britain, Clarendon Press, Oxford, 1981.

Silkin, L. 'Second Reading of the National Parks and Access to the Countryside Bill', Hansard, Vol. 463. 1949

Smith, P.J. The Politics of Physical Resources, Penguin Books, London. 1975.

Smith, S. and Hugget, K. 'Undermined', Farmers Weekly, July, 13th. 1979.

South East Joint Planning team, Strategic Plan for the South East, HMSO, London, 1970.

South Yorkshire County Council, County Minerals Plan : Draft Report of Survey, South Yorkshire County Council, Barnsley, South Yorkshire, 1978.

South West Economic Planning Council, A Strategic Settlement Pattern for the South West, HMSO, London, 1974.

Spooner, D.J. 'Energy and Regional Development', Regional Studies Association Conference Paper, Edinburgh, 1980.

Spooner, D. Mining and Regional Development, Oxford University Press, Oxford, 1981.

Standing Conference on London and South East Regional Planning, Sand and Gravel Extraction the Regional Situation : Policy Suggestions, Standing Conference on London and South East Regional Planning, London. 1974.

Standing Conference on London and South East Regional Planning, Policy Guidelines to Meet the South East Region's Needs for Aggregates in the 1980's, Standing Conference on London and South East Regional Planning, London. 1979.

Stevens, Sir Roger, (Chairman) Planning Control Over Mineral Working, HMSO, London, 1976.

Stoel, T.B. and Scherr, S.J. 'Experience with E.I.A. in the United States', Built Environment, Vol. 4, No. 2. 1978.

Strategic Plan for the North West Joint Planning Team, Strategic Plan for the North West, HMSO, London. 1973.

Tain, P. 'Enforcement in Minerals Planning', Minerals Planning, No.3. 1980, a.

Tain, P. 'A Conditional Response', Minerals Planning, No. 4. 1980, b.

Tandy, C.R.V. 'Industrial Land Use and Dereliction' in Lovejoy, D. (ed), Land Use and Landscape Planning, Leonard Hill, Bucks., 1973.

Tanzer, M. The Race for Resources, Heinemann, London, 1980.

Thorburn, A. 'E.I.A. - The Role of the Planning Authority', Built Environment Vol.4, No.2. 1978.

Tunbridge, I.E. 'Conservation trusts as geographic agents: their impact upon landscape, townscape and land use', Transactions of the Institute of British Geographers, Vol.6. No.1. 1981.

Tyne and Wear County Council, High Spen Area Study, Tyne and Wear County Council, Newcastle upon Tyne, 1975.

Tyne and Wear County Council, Structure Plan, Tyne and Wear County Council, Newcastle upon Tyne, 1979.

Tyne and Wear County Council, Government Changes to the Structure Plan, Tyne and Wear County Council, Newcastle upon Tyne. 1981.

Uden, J. Public Inquiries and the Planning Decision Making Process, University of Glasgow, Glasgow, 1976.

Uvarov, E.B. Dictionary of Science, Penguin Books, London. 1956.

Verney, Sir Ralph, (Chairman) Aggregates : The Way Ahead, HMSO, London, 1976.

Verney, Sir Ralph, 'Planning and Mineral Working', in Planning and Mineral Working, Sweet and Maxwell, London. 1978.

Wallwork, K.L. Derelict Land, David and Charles, Newton Abbot, Devon, 1974.

Warren, K. Mineral Resources, David and Charles, Newton Abbot, Devon, 1973.

Warren, A. and Goldsmith, F.B. (eds) Conservation in Practice, John Wiley and Sons, New York and London. 1974.

Waters, Sir Arnold, (Chairman), Reports of the Advisory Committee on Sand and Gravel, HMSO, London, 1948-1955.

Watson, J.W. North America, its Countries and Regions, Longmans, London, 1963.

West Midland Aggregates Working Party, Regional Commentary, West Midlands Planning Authorities Conference, Birmingham, 1980.

West Midlands County Council, West Midlands County Structure Plan, West Midlands County Council, Birmingham, 1980.

Wiles, J. Reported in O'Callaghan, S, 'Union Slams N.C.B .Report on Belvoir Subsidence', Farmers Weekly, August, 17. 1979.

Winward, J. 'Half-time at the Vale of Belvoir', Town and Country Planning, Vol, 49. No. 3. 1980.

Yorkshire and Humberside Economic Planning Council, Regional Strategy Review 1975. HMSO, London. 1975.

Zimmerman, E.W. World Resources and Industries, New York. 1951.

Zuckerman, Lord, (Chairman), Report of the Commission on Mining and the Environment, London, 1972.

Index

Addison Committee, 20
Adit Mining, 16
Aggregates, 3, 28 – 31, 35 – 37, 55, 61
Aggregate Working : National Planning Guidelines, 6, 91, 101, 111
Agricultural land, 52 – 53, 54, 58, 80
Alternatives, 34, 42 – 45, 115
Amenity, 5, 7, 9, 38, 58 – 62, 74 – 75, 92, 107 – 108
Amey Roadstone Corporation, 71
Appalachia, 119
Archaeological sites, 61, 107

Bedfordshire, 14, 69 – 70
Bell pits, 16
Belvoir, Vale of, 9, 54, 56 – 57, 61, 69, 94, 101, 117, 129
Brick making, 14
British Aluminium, Invergordon, 4
Brick Development Association, 134, 145
British Gas, 137
British Industrial Sand, 126, 141
British National Oil Corporation, 137
Butterwell, Northumberland, 9, 56

Caernarvonshire, 45
Canada,   15
Central Planning Authority,   20
Cheshire,   91
Coal mining, 3, 14, 17, 34, 45, 56 – 58, 70 – 71, 88, 90, 105
Colliery spoil,   43
Common Agricultural Policy, 140
Confederation of British Industry, 122, 126
Consultation areas, 71, 91, 92
Control of Mineral Working, 5, 77 – 78, 80 – 81, 84, 101, 103, 106, 143, 144,
                                                        145
Corby,   14, 17,
Cornwall, 90 – 91, 148
Council for Environmental Quality, 131, 132
Council for the Protection of Rural England, 20
Countryside Commission,   9
Cycle of Development, 24, 45, 111